Essentials of

ECONOMIC DECISION ANALYSIS

for Chemical Engineering

Ralph W. Pike

Printed by CreateSpace, An Amazon.com Company

Ralph W. Pike
Louisiana State University
Baton Rouge, LA 70803

Disclaimer

ISBN-13: 978-1507771723

ISBN-10: 150777172X

Dedication

To my wife, Patricia, son Brian, and daughter, Charlene

Table of Contents

4

Preface

There are things that chemical engineers need to know and things that they need to know about. They need to know the time value of money, and they need to know about using Solver in Excel for process optimization problems. They need to know how to describe a process or plant in a flowsheeting program, and they need to know about the options for the thermodynamic packages in the flowsheeting program. They need to know how to do material and energy balances on chemical processes, and they need to know about how to incorporate rate equations and equilibrium relations in the material and energy balance equations. The topics in economic decision analysis given here have details to ensure the depth for need to know and the breadth for the topics that are in the need to know about category.

The topics follow in order of developing the information needed for economic decision analysis. Beginning with market analysis, this leads to the decision to evaluate a new process or plant. Then several levels of analysis are used to determine cost of manufacturing a new product to compare with the sales estimated from the market analysis. Profitability analysis uses the net present value, rate of return and the economic price for decisions.

Total cost assessment used the triple bottom line to incorporate economic, environmental and sustainability costs and sustainability credits in the decision analysis. The optimal configuration of plants in a chemical production complex was evaluated using global optimization to the determination the best value of triple bottom line. Results included using multiobjective optimization to determine Pareto optimal solutions. Monte Carlo simulation was used to determine the sensitive of the optimal solution to the parameters in the constraint equations.

Three appendices are included. In Appendix A, typical estimating charts are given for individual equipment cost and plant cost for use with quick estimation of total plant costs to be used with total product costs for preliminary estimates of profitability. In Appendix B supply and demand elasticity are described with evaluations for corn as a feed and a fuel, illustrating the difficulty in finding data to do the evaluation of these frequently used concepts. In Appendix C the latest status on global optimization is summarized with the currently more successful algorithms available in GAMS.

Comments, recommendations and suggestions for improvement are welcome; email pike@lsu.edu.

Ralph W. Pike
Louisiana State University
Baton Rouge, LA
February 18, 2015

6

Motivation

It is important to understand and appreciate the concepts of economic principles involved in engineering decisions to invest corporate funds for the return to investors. Methods for economic decision modeling are described and the objectives are to understand:

Importance of forecasting revenues from the sale of products

Methods for estimating plant and manufacturing costs

Several methods used to evaluate plant profitability

Analysis and optimization of chemical production complexes

Introduction

Money invested generates more money, and economic decision analysis gives ways to make systematic evaluations of alternatives for investments. These economic decisions allocate the corporation's capital for a maximum return on these investments. An economic evaluation of a proposed investment includes determining the expected profit and capital expenditures, and these investments can include construction of a new plant and expansion of existing facilities, among others. The decisions are based on reports from forecasting revenues from the potential sales of the plant's products.

New plants, products and technology require new capital, and most firms have limited resources. Consequently, investment decisions require capital budgeting. This plant profitability analysis is the evaluation and selection of the best investments from a set of alternatives. Methods for evaluating investments include net present value and rate of return, among others, for private companies and benefit-cost ratio for public works projects. All of these come under the purview of plant profitability analysis.

Risk is a part of the decision process, also. The analysis of projects must incorporate the level of risk to be able to judge projects with high returns and high risks with those having lower returns and more certain outcomes.

A company typically has several projects competing for funds to be invested. The projects are ranked based on their net present value and risk. This is an economic decision problem. However, each project is an optimization problem in itself. For a valid comparison among projects, the optimum design is required for each project to have the maximum net present value.

Before developing economic and process models, it is important to understand optimization method capabilities. Optimization methods take advantage of the mathematical

7

structure of the economic and process models to locate the optimum, possibly a global optimum. Models should be developed to use these capabilities to locate optima. For example, if linear equations can provide satisfactory representations of the economics and processes of the plant, then linear programming can be used to locate the global optimum. However, if the models are nonlinear, optimization methods can only guarantee a better point than the starting point, possibly a local optimum.

Some terms used in economic decision analysis are given in Table 1 (Perry's, 1997), and using an annual basis is standard. Sales (S), the income or revenues received from customers who will purchase the plant's products and byproducts, is estimated using market research. Total product costs or total annual expenses of a plant (C_T), the sum of manufacturing costs (C_M) and general expenses (C_G), are estimated by cost engineering using a flowsheeting program and related information. Capital expenditure annually (C_{cap}) include funds for working capital and plant additions and modifications.

The annual gross profit (P_G) is the sales less manufacturing costs and depreciation. Depreciation is a business expense. The annual net profit (P_{net}) is the gross profit less the general expenses. The annual net profit after taxes (P_{xt}) is the net income after taxes less depreciation. Methods for depreciation are given later in the chapter.

Referring to Table 1, the net or cash annual income before taxes (I_{net}) is the income from sales less total product costs. Plant equipment can be depreciated, and the taxable income is the net income less depreciation. Taxes (t) are paid on a sliding scale, and a corporate rate of 35% is used for estimation in the U. S. and approximately 50% in developed countries. Then the net income after taxes (I_{xt}) can be determined.

In profitability assessments, annual cash flows are more meaningful than net profit. The net annual cash flow after taxes (CF_{xt}) is the net profit after taxes less the annual expenditure of capital for additions and modifications. Net annual cash flows are used in discounted cash flow calculations to determine the net present value and the rate of return that are two key measures used in economic decision analysis. These terms are discussed in detail, and the following example illustrates these calculations for a preliminary design of the aniline process.

Economic decision analysis provides the framework for economic feasibility studies as part of a company's on-going planning process. In the next section the formulation of economic decision analysis is described for plant design and for operating plants. This is followed by a section that describes market analysis and forecasting revenues. Then methods to estimate plant equipment and manufacturing costs are described. These sales and costs are combined using profitability analysis for economic decisions.

Table 1. Terms Used in Economic Decision Analysis on an Annual Basis, after Perry's, 1997.

Sales (sales price S_p x product flow rate-mass per yr. m)	$S = S_p\ m$
Manufacturing costs (see Table 3)	C_M
General expenses (see Table 3)	C_G
Total product cost (total annual expenses or manufacturing cost)	$C_T = C_M + C_G$
Purchased equipment cost	$C_{purchase}$
Installed equipment cost (See Table 2) or fixed capital investment, FCI	$C_{installed} \sim (2.5\ to\ 6.8)\ C_{purchase}$
Total plant cost (See Table 4) or total capital investiment, TCI	$C_{total\ plant} \sim 2.4 C_{installed} = CF_0$
Capital expenditure annually	C_{cap}
Depreciation and allowance for tax purposes	$D \sim C_{installed}/$economic life
Gross profit	$P_G = S\ -\ C_T - D$
Net annual income before taxes	$I_{net} = S - C_T$
Net annual profit before taxes	$P_{net} =\ P_G - C_G$
Net annual cash flow before taxes	$CF = I_{net} - C_{cap}$
Taxable income	$I_t = I_{net} - D$
Taxes (tax rate t $\sim$ 35% of taxable income in U. S.)	$T = t\ (I_{net} - D) = t\ I_t$
Net annual income after taxes	$I_{xt} =\ I_{net} - T$
Net annual profit after taxes	$P_{xt} =\ I_{xt} - D$
Net annual cash flow after taxes	$CF_{xt} = I_{xt} - Ccap$
Value Added (sales - raw materials cost - utilities)	$P_{value\ added} = S - C_{raw\ matls} - C_{utilities}$
Earnings - net annual income after taxes from continuous operation excluding significant extraordinary and nonrecurring items	$E = I_{xt}$
Profit Margin - after tax earnings as a percentage of sales	$P_{margin} = P_{xt} \bullet 100/S$

Example 1: Economic Analysis of an Aniline Plant Preliminary Design

A preliminary design of a plant to produce 100 million pounds per year of aniline from the reaction of phenol and ammonia has been completed. The following information has been developed.

Plant capacity	100 million pounds/yr.
Plant installed cost	$6.0 million
Total plant cost	$11.9 million
Total product cost	$46.3 million/yr
Annual capital expenditures for worn out equipment	$0.5 million/yr
Estimated annual sales	$53.2 million/yr
Economic life	10 yrs
Tax rate	35%
Minimum attractive rate of return	15%

Depreciation: straight-line method with no salvage value.

Performing the following economic analysis, determines:

Net annual income before taxes = sales - total product cost = $53.2 - $46.3 = $6.9 million

Net annual cash flow before taxes = net annual income before taxes - annual capital expenditures
for worn out equipment
= $6.9 -$0.5 = $6.4 million/yr

Depreciation = plant installed cost/economic life = $6.0/10yrs. = $0.6/yr.

Taxes = tax rate*taxable income = tax rate*(net annual income before taxes- depreciation)
= 0.35(6.9 - 0.6) = $ 2.2 million

Net annual income after taxes = net annual income before taxes – taxes

= $6.9 - $2.2 = $4.7 million

Net annual cash flow after taxes = net annual income after taxes - annual capital expenditures for
worn out equipment

= $4.7 - $ 0.5 = $4.2 million

Net present value (NPV) based on the net annual cash flow after taxes and minimum attractive rate of return [See Equation (24).]

$$\text{NPV} = -CF_0 + A\{[(1 - (1+i)^{-n}]/i\} = -\$11.9 + \$4.2\{[1 - (1.15)^{-10}]/0.15\} = \$9.2 \text{ million}$$

Rate of return where the net present value is zero [See Equation (25).]

$$\text{NPV} = -CF_0 + A\{[(1 - (1+i)^{-n}]/i\} = 0 = -\$11.9 + \$4.2\{[1 - (1+i)^{-10}]/i\} = 0, \text{ solving for } i = 33.3\%$$

Economic price [See Equations (21) and (28).]

$$\text{Annual cost of capital} = \text{EUAC} = P*(A/P) = P*\{i/[1 - (1+i)^{-n}]\} = \$11.9\{0.15[1 - (1.15)^{-10}]\}$$

$$= \$2.37 \text{ million/yr.}$$

Economic price = (total product cost + annual cost of capital + C_{cap})/ product rate

$$= (\$46.3 + \$2.37 + 0.5) \text{ million/yr/100 million pounds/yr} = \$0.49/\text{lb}$$

Overview of Plant Design and Operating Plant Optimization

There are two types of industrial optimization problems: design optimization (new plant, plant expansion, debottlenecking, adding new technology) and operations optimization (process, plant, multi-plant) as shown in Figure 1. Both require economic models that describe the profit to be maximized or the cost to be minimized, and a plant simulation (process model) that is used to predict the performance of the plant.

As shown in Figure 1, the economic model for optimal plant design is net present value, while the economic model for optimal plant operations is net profit. Net present value is the annual cash flows discounted to the present value after all the capital and operating expenses have been paid. To compute the net present value for design optimization, the annual net profit has to be estimated over the life of the project. This requires the estimation of the sales price and demand for the product and of the capital and total product cost, manufacturing costs or total annual expenses. Net profit before taxes is the difference between funds received from selling the product and the total product cost or manufacturing costs (raw material and operating costs, taxes, administrative expenses, etc.). See Table 1. The net profit is determined annually in operations optimization.

```
                      DESIGN              OPERATIONS

Economic Model
                net present value         net profit

Constraints
                plant configuration       capacities of process
                                          units

                    material and energy balances
                    availability of raw materials
                    demand for product

Results
                capacities of process units    operating conditions
                and operating conditions


Process model from the plant design is used for the simulation of the
operating plant

Economic data estimated in plant design is replaced by the current
data


        Figure 1 Comparison of Design and Operations Optimization
```

To evaluate the net present value required forecasting revenues and estimating capital and manufacturing costs. Forecasting revenues from sales prices and demand for product requires market research, and capital and manufacturing cost estimation requires cost engineering. Sales prices are determined by market conditions, including the demand for product and its availability from competitors. They are not related to the economic or manufacturing price that includes a minimum attractive rate of return (MARR). Income for a new project, net profit, requires estimations of the quantity of product that can be sold free on board (FOB), the sales prices, product specifications, and any freight allowances and other charges.

Referring to Figure 1, the constraints are the plant simulation. This simulation includes equalities such as material and energy balances, rate equations, and equilibrium relations, and inequalities such as capacities of process units, demand for products, and availability of raw materials. These equations provide the constraint equations for optimization. The process and economic models are used with an optimization algorithm to determine the best design and operating conditions that maximize the profit or minimize the cost of operation.

Optimal Plant Design: Design optimization determines the best operating conditions and capacities of the process units from the plant simulation. Flowsheeting programs (HYSIS, ASPEN PLUS, PRO II) are standard tools employed for preparing the design, and they include an optimization capability with several algorithms. In order to complete material and energy balances and determine process unit capacities, a process configuration and stream specifications are required as illustrated in Figure 2. An economic model is required, and it is usually the net present value. Also, the profit margin can be used for screening purposes, and it is the difference between the sales revenues and cost of raw materials. A nonlinear programming algorithm is used by a flowsheeting program to calculate unit sizes and flows that minimize the costs. However, the best plant configuration must be determined by a case study that yields only the best among the cases proposed.

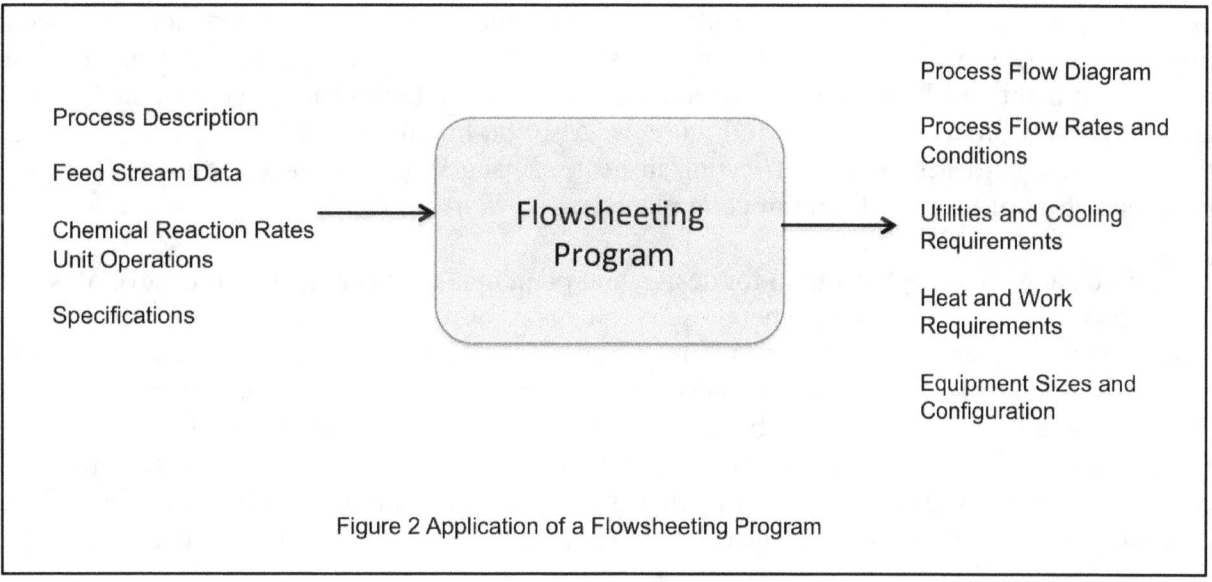

Figure 2 Application of a Flowsheeting Program

The optimal plant configuration problem is one where a superstructure containing a number of comparable units is proposed for the plant, and the best arrangement of units is determined using an optimization algorithm. This is a mixed integer nonlinear programming problem (MINLP), and there are optimization procedures available to solve the MINLP. However, the solution is more difficult, and the methods are more elaborate than flowsheeting optimization. This is an important area for current research, and the equivalent of the Kuhn-Tucker conditions do not exist for MINLP. In Appendix C, commercial programs available to solve the optimal plant configuration problem are summarized

Optimal Plant Operations: Plant operations optimization determines the best operating conditions that maximize net profit from plant operations. The objective has to satisfy plant material and energy balances, process unit capacities, availability of raw materials, and demand for products. The net profit is the difference between the funds received from selling the product

13

and the manufacturing costs. The manufacturing costs include operating and raw material costs, taxes, administration, and other costs. (See Table 1) A simpler version of the net profit is the "value added" economic model that is the difference between the sales and raw material where operating costs and all other costs are assumed constant. It is expected that the net profit after taxes will be comparable to the estimates of the cash flows made for this time period when the plant was designed and that the plant will generate the anticipated return on the investment.

Unlike in the design optimization problem, the plant configuration is specified, thus making the operation optimization problem somewhat easier. However, there are multiple levels of optimization that must be considered as shown in Figure 3. One level is the optimal scheduling problem of corporate headquarters to distribute raw materials among the company's plants to maximize profits in producing, transporting, and marketing products to consumers worldwide. Also included is the optimal scheduling problem of the individual plant to set operating conditions to produce required products from allocated raw materials for a maximum net profit or minimum cost of operations. The best schedule is determined for steady-state daily or weekly average flow rates for the plant. Finally, there is on-line optimization of process operations to determine the set-points for the distributed control system of the individual process units in the plant which give the best operating conditions while producing the specified quality and quantity of products as shown in Figure 4. Also, on-line optimization keeps track of such things as catalyst deactivation and scaling in heat exchangers by parameter adjustments in the process models of the units from sampling plant data.

To summarize, optimization for design and plant operations are different in several ways. The economic model for design is net present value and for operations is net profit. The process model for operations includes the plant configuration, material and energy balances, availability of raw materials, and demand for products. The process model for design plant does not have a plant configuration, and it has to be determined, along with the capacities of process units. Finally, design optimization determines the capacities of individual units and plant operating conditions. The process model from plant design can be transferred to the simulation of the operation of the plant. Economic data estimated in plant design is replaced by actual data.

Energy Integration (Pinch Analysis) Energy integration evaluates the heating and cooling requirements for a process or plant. The plant's heat exchanger network (HEN) is used to determine the amount of heat transferred among the hot and cold streams and the additional need for utilities (e.g., steam and cooling water). The objective is to minimize the required utilities knowing each stream's inlet and outlet temperatures, mass flow rates and physical properties. Energy is said to cascade down from the highest temperature and up from the lowest temperature. In this process of matching hot and cold streams, a temperature is found, the pinch temperature, where a hot utility must be supplied above this temperature and a cold utility must be supplied below this temperature. An excellent description is given of the procedure and extensions by Knopf, 2012.

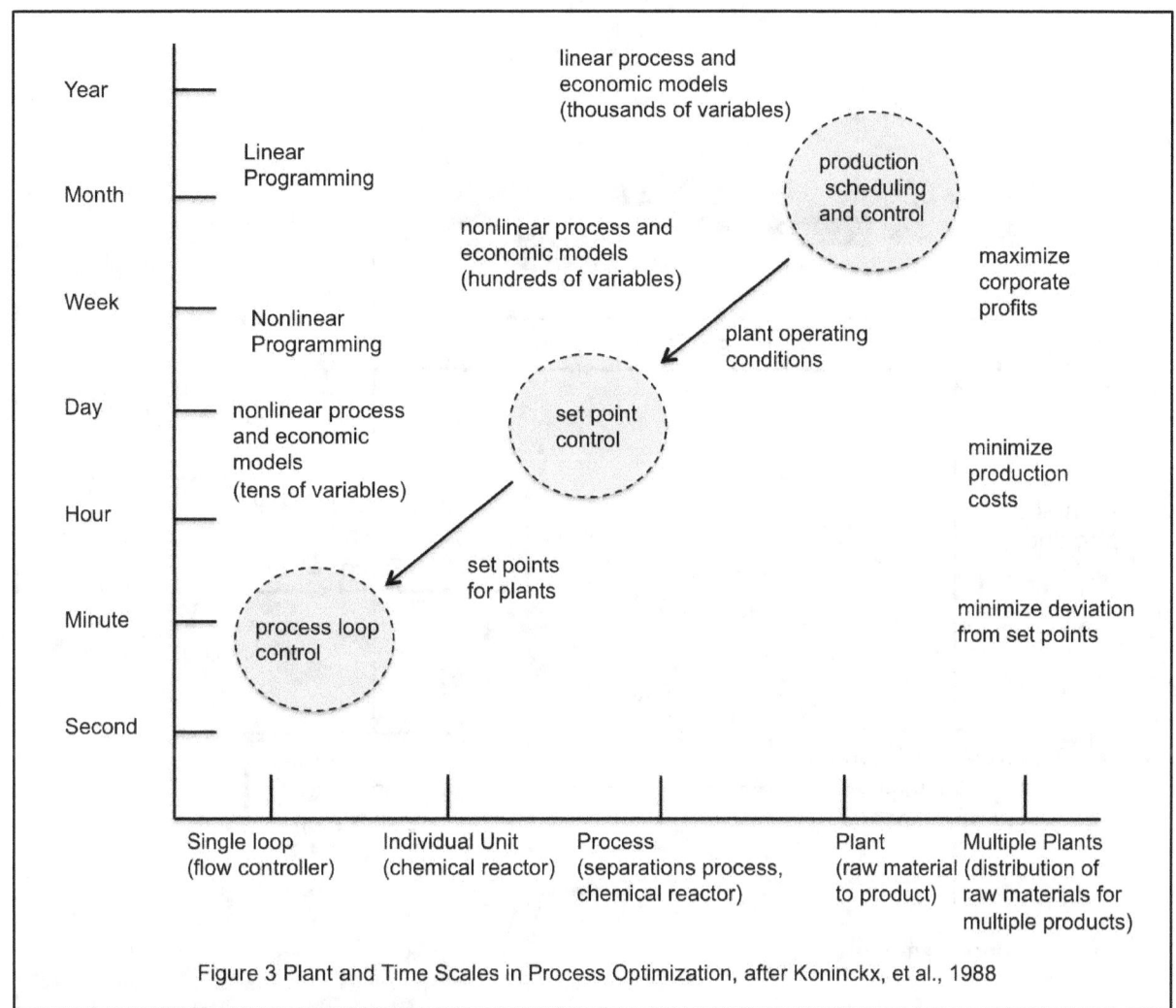

Figure 3 Plant and Time Scales in Process Optimization, after Koninckx, et al., 1988

Heat exchanger network optimization can be solved as a mixed integer linear programming problem or using heuristic rules. The methodology has evolved into a general method for energy and mass integration, and it has been applied to mass exchangers, especially process water, chemical reactor networks, and batch process site integration. The heat exchanger network optimization has been successfully applied to existing operating plants, and there are several commercial computer programs available to perform this optimization.

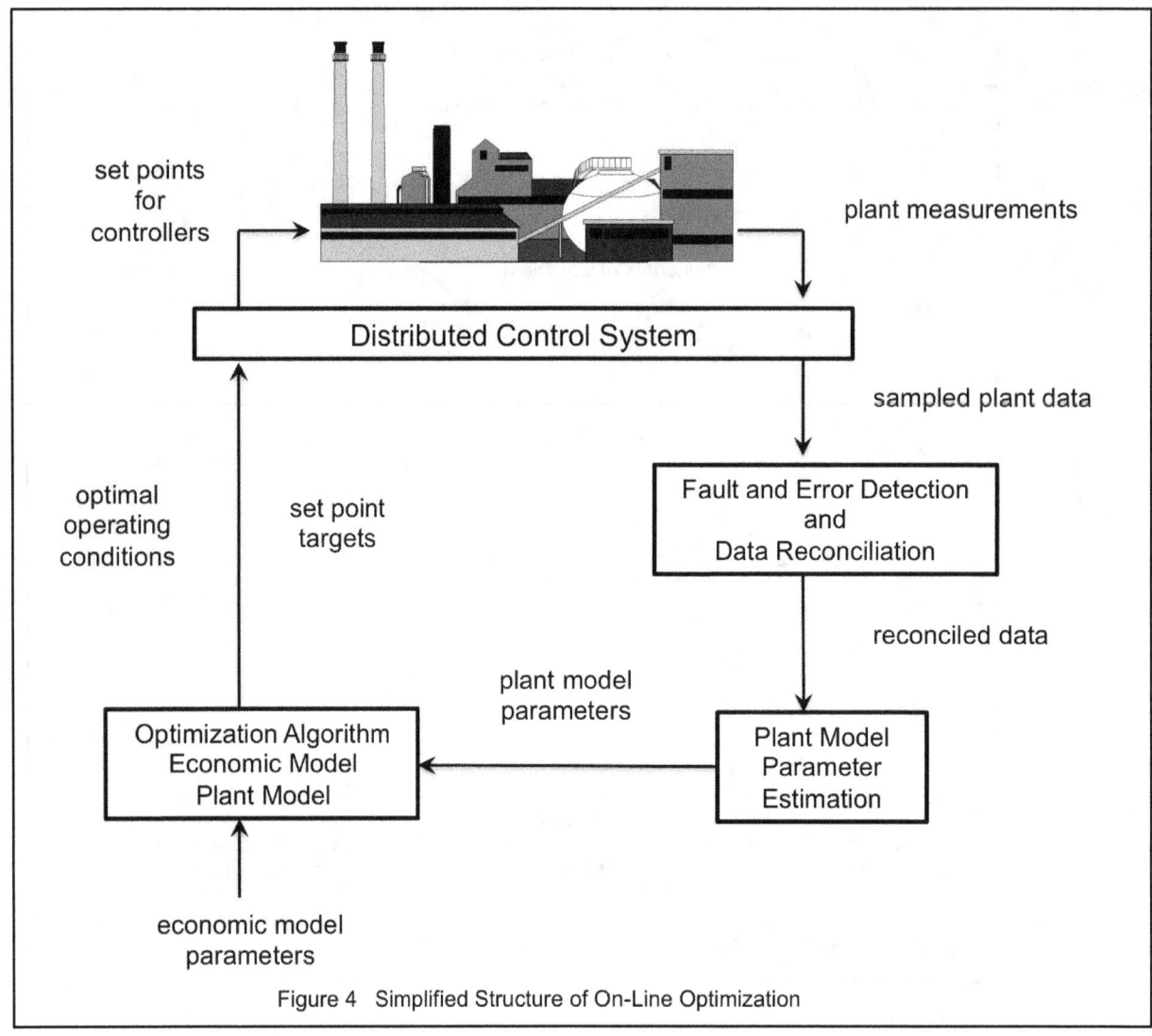

set points for controllers

plant measurements

Distributed Control System

sampled plant data

optimal operating conditions

set point targets

Fault and Error Detection and Data Reconciliation

reconciled data

plant model parameters

Optimization Algorithm
Economic Model
Plant Model

Plant Model Parameter Estimation

economic model parameters

Figure 4   Simplified Structure of On-Line Optimization

Decision Analysis for Commodity and Specialty Chemicals: Distinctions are made between design and analysis for commodity and specialty chemicals production. Typically, commodity chemicals are made in large-scale chemical manufacturing sites around the world to satisfy global markets. At these locations there are interrelated chemical plants that use the infrastructure such as utilities (power, steam, cooling water, etc.), port facilities, and road and rail terminals. Houston, Texas, the Antwerp port area in Belgium and Ras El Anouf, Libya are among the largest sites (Xu, 2004). Specialty chemicals are usually manufactured in relatively small-scale chemical plants, many times using batch-processing techniques. Most specialty chemicals are organic chemicals that are used in a wide range of products used by consumers and industry. Specialty manufacturing units are flexible because the products, raw materials, processes and operating conditions, and equipment mix may change on a regular basis to respond to the needs of customers (Cussler and Moggridge, 2001). Products include pharmaceuticals, polymers, electronic chemicals and industrial gases to mention a few. The methodology and

16

tools of economic decision analysis are the same for both commodity and specialty chemicals, including revenue forecasting, estimating plant and manufacturing costs, evaluating profitability and optimization.

Market Analysis - Forecasting Revenues

The demand for chemicals is strongly influenced by what is happening in the economy, in general. The chemical industry is cyclic as the demand for products oscillates with uses in the housing, automotive and consumer products and with new plants coming on-stream. Also, new products are continually being developed from basic and applied research, and products go through life cycles. For example, celluloid was one of the first plastics, and others have replaced it that have more desirable properties. Another example is synthetic textile fibers that grew at a rate of 20% per year from 1950 to 1970 that is comparable to the software industry in the 1990's (Cussler, 1999). Then from 1970 to 1990 it grew at the rate of the overall economy, and the industry remained profitable by building larger, more efficient plants and consolidating production. In the 1990's profitability has continued by restructuring and expansion into biotechnology.

Future changes in the chemical industry are planned to have less emphasis on commodity chemicals that have small profit margins, and to expand specialty chemicals and products from biotechnology and nanotechnology that have large profit margins. This means that there is increasing emphasis on product design that has the following four steps.

Identify Customer Needs - Market Research

Generate Products to Meet Needs - Research and Development

Select Among Products for Development - Economic Feasibility Evaluation

Determine Processes to Manufacture Products - Process Design

A company must continually develop new products since patent protection is for 17 years. For example, the patent on Monsanto's "cash cow" Roundup, a potent herbicide that accounted for 17 percent of Monsanto's total annual sales of $9 billion, expired 2000; and Monsanto's NutraSweet patent expired in 1992. In addition, existing products may have to be replaced with new ones as a result of government regulations. For example, tetraethyl lead and chlorinated fluorocarbons are no longer manufactured. The world demand for fluorocarbons, high margin chemicals for refrigerants, was over 6000,000 tons per year in the mid 1980's (DeSimone and Popoff, 1997), and they are no longer produced.

Revenue as a Function of Price per Unit and Demand: The sales price of a product depends on the demand for the product and its availability from competitors. It is not related to the manufacturing costs. If a company were to market a unique product, then it would be reasonable to assume that in the absence of competition and that demand exists, the company could selling the product at a relatively high price. An example would be a prescription medicine that is covered by a patent. Although the unit manufacturing cost of the medicine could be in cents per dose, it could be sold for several dollars per dose. The price is usually set by the "customers' ability to pay" rather than the manufacturing cost, and research and development costs are said to be recovered. Medicines are indispensable, and a steady demand for such products always exists. The same is true for energy, and the price of crude oil over a period from 1974 to 2014 is shown in Figure 5. The data in Figure 5 is for the monthly average composite refinery acquisition cost (composite price) reported by the U. S. Energy Information Agency. The composite price is the combination of the price paid by refineries for domestic crude and imported crude. In the 1970's, domestic crude price was less than imported crude price by about $5.00/bbl or less, and recently domestic crude price was greater than imported crude price by about $5.00/bbl or less according to the EAI Monthly Energy Review, September 2014.

In order to maximize profit, companies set the sales price in order to capture as much of the market as possible. Consequently, before any new project is undertaken, the projected income is determined by estimating the quantity of the product that can be sold, the product specification, its sales price, etc. Such information is obtained by conducting studies on the existing and potential markets by means of market research. In this section we will elaborate on the concept of profit as a function of unit price and demand and briefly discuss some of the methods of analysis used in market research.

As mentioned before, the objective of any venture is to make a profit. Exactly how much profit can be made depends on the market demand for the product, which invariably depends on the unit price of the product. If the price of the product increases, then the demand decreases (other things being equal), and vice versa. This change in demand with price is known as "price elasticity of demand," and any company that wishes to market a new product or expand sales on an existing one must have some knowledge of the price elasticity of demand for the product.

Market conditions are carefully reviewed in order to arrive at an acceptable sales price of new products. For instance, if a new product is to compete with other similar products already on the market, the pricing should be such that it offers some obvious economic advantage to the customer. An example from Valle-Riestra, 1983, illustrates this point by considering a new specific herbicide for weed control in wheat fields that is five times as effective per kilo as the nearest competitor. However, the price of this new product cannot be five times as high, because, all other things being equal, the customer does not gain that much economic benefit. Instead, if the price is set to twice that of the competition, it might provide a strong incentive for customers to switch

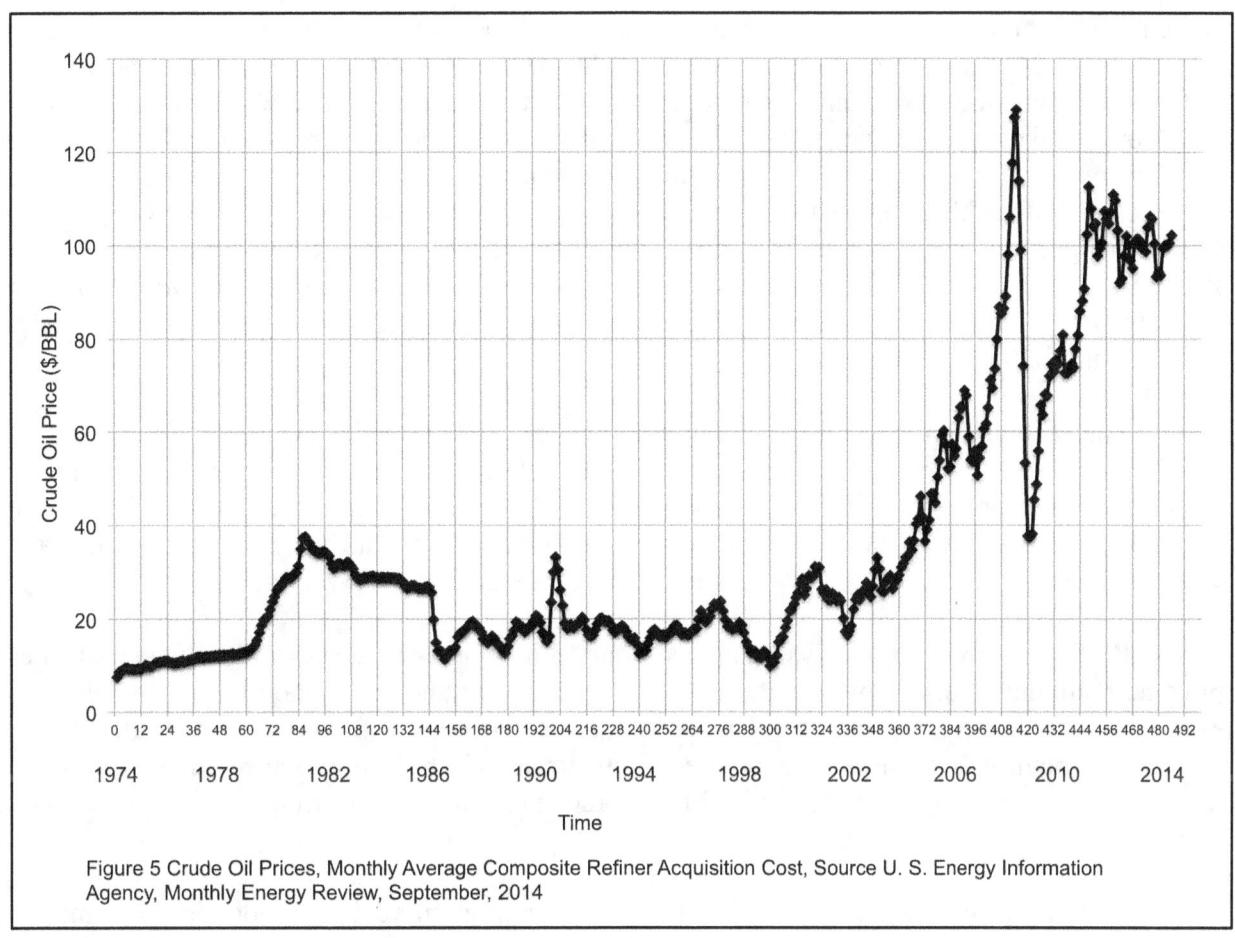

Figure 5 Crude Oil Prices, Monthly Average Composite Refiner Acquisition Cost, Source U. S. Energy Information Agency, Monthly Energy Review, September, 2014

Market Research: Market research is used to study existing and potential markets to provide the information needed to estimate the market penetration of a product (the percentage of the market that a product can obtain from a new plant). Investment in a new production facility must be justified on the demand for the product. However, a rule of thumb is that a product from a new plant can be expected to obtain no more than 10% of the market. Large companies have market research groups of specialists who gather and project sales information that is available on each new or changed product under consideration. Also, they make recommendations on new plants, product quality, and specifications including required packaging, sales prices, discounts, freight equalization, and marketing methods.

There are systematic methodologies used in market research, but any information that yields the required answers is helpful and is utilized. A market survey is an essential method of market research, and they are routinely conducted. Market surveys are done in three steps. First, one must evaluate information available on products. This includes amount sold, growth patterns, the industries that are using the product, the end users, the sizes and locations of competitors' plants, price history and trends, shipment methods and containers, specifications,

19

and environmental impact. This information can be obtained through various sources such as U. S. Department of Commerce surveys, *The Wall Street Journal*, annual reports and income tax statements of chemical companies, bulletins from financial analysts, "*C & E News*" and similar magazines, and consulting companies. Information on current price of chemicals can be found in the *Chemical Marketing Reporter* (www.chemexpo.com/cmronline) and *Chemical Weekly* (www.chemicalweekly.com). Also, web sites that contain price information include Chem Expo (www.chemexpo.com) and Chem Industry (www.chemindustry.com). There are companies that specialize in providing chemical market information such as ICIS (*www.icis.com*) for petrochemicals, energy and fertilizer. Additional information about on-line chemical sales is given in later in this section.

The second step is to interview potential customers for purchasing needs. This involves field calls by sales representatives and asking potential and existing customers the same set of questions, either orally or in the form of a questionnaire. Some of the questions could be about their likes, dislikes, expectations, the amount of money they would be willing to spend, and when, how much and under what conditions would they buy a specific product.

Finally, a sensitivity analysis must be done to include estimates of the influence of sales price and amount sold. Below a certain capacity a plant may not be economically feasible. An example is ethylene plants that must have very large capacities for an economy of scale. When an ethylene plant is brought on-stream, there is a surplus of ethylene in that region of the world until demand catches up with supply, and it becomes profitable to construct another world-class plant.

In addition to the above-mentioned steps for market surveying, production and market data are evaluated; and extrapolations are done using deterministic and statistical models including the effect of inflation. Estimation of market and production data uses forecasting models. A forecasting model (or scenario) can be one of three types: explanatory or casual models, time series or trend models, and judgmental or scenario forecasts (Cassinatis, 1988). Casual models are based on regression equations where the forecaster postulates cause and effect relationships to fit available data using the idea that one event is associated with another event. Time series or trend models are regression equations with time as the independent variable. These models use linear and nonlinear extrapolation to predict trajectories of price and demand. Also, they use concepts such as persistence prediction, which states that trends that happened in the past will continue in the future. Judgment or scenario forecasts are based on the opinions of experts about the future direction and level of sales. In summary, selection of forecasting methods depends on the objective of the forecast, availability of data, and the time horizon for the forecast.

Forecasts can provide a fairly accurate picture of the sales potential for well-established commodity products. Growth rate is approximately the same as the entire industry. For newer or specialty products, there is less data available, and sales and customer acceptance are more uncertain. These products can enter the commodity phase with a growth that is higher than the industry average. Risk is lower in estimating the amount and the fraction of the total market for a product that can be sold from a new plant when it is in or entering the commodity phase. Ten percent of the market is usually the maximum for an established product from a new plant unless there are some unusually favorable conditions. However, all products go through a life cycle.

Forecasting typically uses linear or quadratic fits to historical data for product demand and sales price, but this has limited success. It is often hard to predict the development of new markets and new technology, the shortage of supply, the raising of prices by the main supplier, and the appearance of superior products at competitive prices. Forecasting any events beyond five years is usually not attempted.

Trend curves of similar products are used to predict demand for the new product. Examples of trend curves are shown in Figure 6 and 7. This type of information is used to estimate new plant capacity, both the initial size and the size and timing of its later increments from increased demand. The range of the forecasting should cover the plant's capacity to meet current demand to its capacity to fulfill ultimate demand. This forecast requires a model of demand, production unit economic life, and other factors, including the life cycle of the product. Also, these are optimized over time to determine additional capacity increments, if possible. For example, when Aramco installed basic petrochemical production capacity in Saudi Arabia in the 1970's, these products were shipped to Europe; and this caused a shut-down in plant capacity in Europe. However, demand grew in the late 1980's, and the European plants were back running at full capacity. During this time period the average of all petrochemical prices dropped by 25% over a six-year period while production rate doubled.

In addition to market research, a company has to have a marketing strategy. This is typically a product focused or a customer focused strategy, Gizinski, 2000. The marketing department is challenged with how to sell more (product focused) or what do customers want and how can we serve them (customer focused). Expanding sales can include selling more products, adding product lines or services acquiring another business entity and entering different markets. Selling more of an existing product usually means displacing competitors' sales. Applying market research includes determining potential customers' beliefs about their requirements, analyzing sales of new customer classes and determining adjustments to be made to meet customer needs. This effort includes interviewing manufacturing, purchasing and others who influence buying decisions for potential customers, and each will voice different requirements. If the marketing department is successful in determining customers' needs and the market potential for additional sales appears attractive, then financial estimates are made for product development, manufacturing adjustments, and sales force and technical support training. If this evaluation gives an acceptable return on investment (capital employed), then marketing's four P's are applied - product, price, packaging and promotion. Industrial companies focus much of their marketing effort on product and price. The economic or manufacturing price establishes a pricing "floor." For customer-focused companies, the "packaging" goes beyond specifying a container, e.g., tank car, bag, bucket, or dispenser. It includes delivery mode and service support. The marketing department offers the customer a "package" that includes product, service and practical technical support.

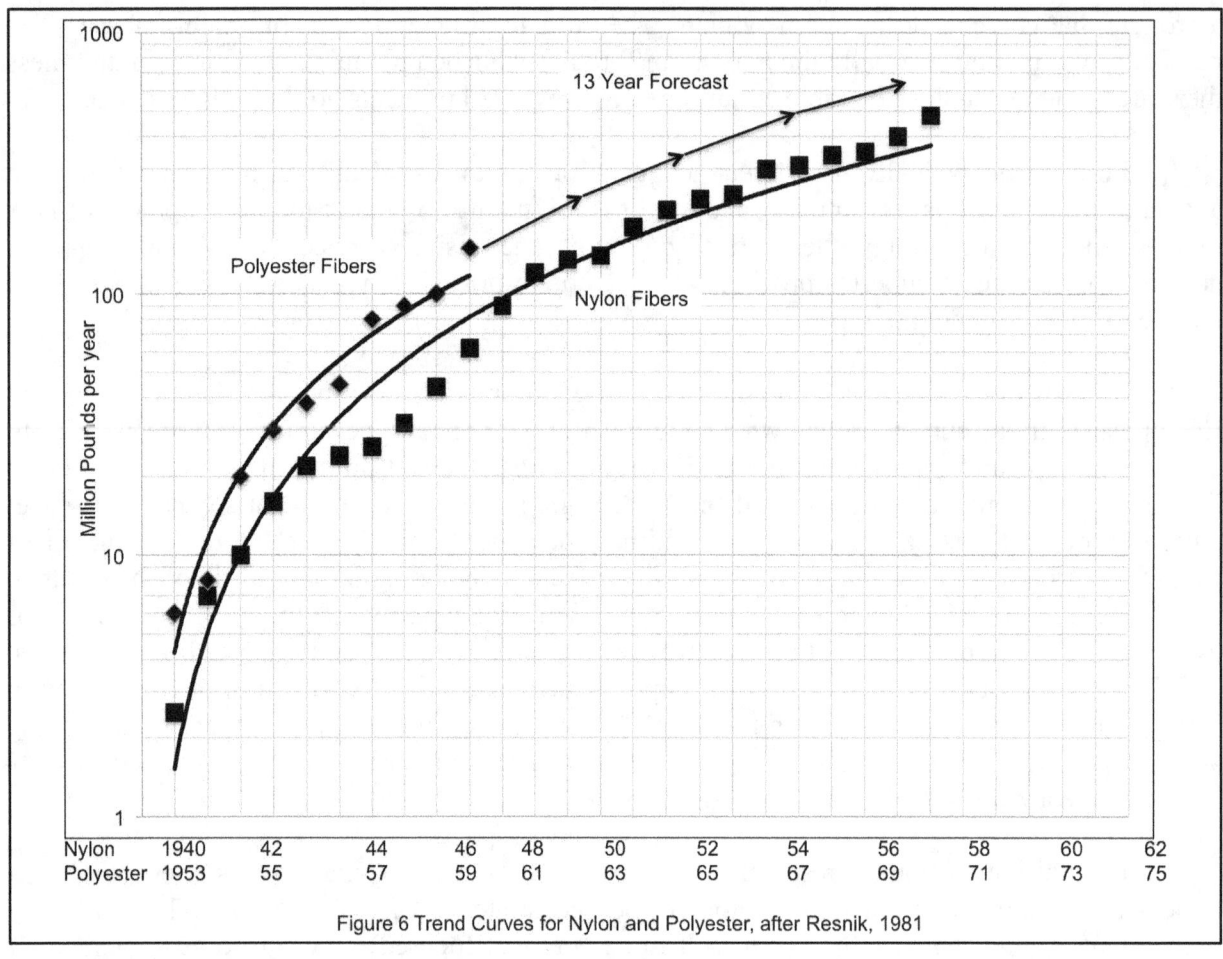

Figure 6 Trend Curves for Nylon and Polyester, after Resnik, 1981

Promotion for existing customers includes literature with detailed technical product descriptions. To develop new customers, promotions address the how the supplier will provide the most benefits per dollar purchased and details about the product and supporting service. Gizinski, 2000 provides illustrations of these methods from the epoxy adhesive business.

On-line chemical exchanges have grown rapidly. Their business models include a general exchange, a commodity trading area, private auctions and a hub for systems transactions. Transaction volumes in 2001 were over $3.2 billion. Leading companies include ChemConnect (www.chemconnect.com), ChemPoint (www.chempoint.com), Chem Industry (www.chemindustry.com), Elemica (www.elemica.com) ChemCentral Online (ww.chemcentral.com), cc-chemplorer (cc-chemplorer.com) and Chemxl (www.chemxl.com). These companies are dynamic; and mergers, startups and dropouts are frequent. The customers and investors in these Internet organizations are the major chemical companies, Thayer, 2001.

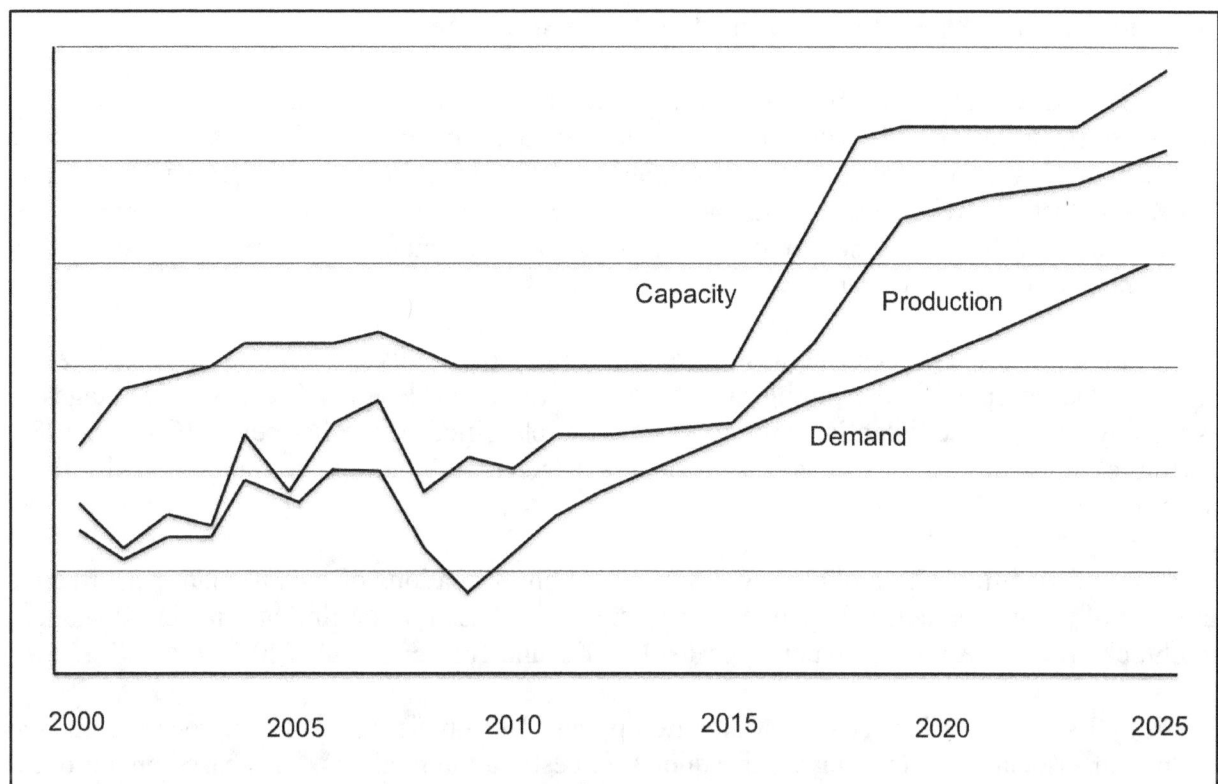

Figure 7 Trend Curves for High Density Polypropylene (HDPE) source ICIS World Plastics Annual Study brochure http://img.en25.com/Web/ICIS/FC0111_CHEM_201210.pdf accessed 10-16-2014

Supply and Demand: In Appendix B, Supply, Demand and Price Elasticity, price elasticity of demand (PED) and price elasticity of supply (PES) are presented with application to corn as a feed and fuel. Describing supply and demand of goods and services in the economy is complicated. To provide insight, two measured parameters are used: price elasticity of demand and price elasticity of supply. In economics, elasticity is defined as the ratio of the percent change in one variable to the percent change in another variable.

Price elasticity of demand (PED) is a measure of the responsiveness of quantity demanded to changes in price. Mathematically, it is the ratio of percent change in a quantity of goods or services, Q, to the percent change in price, P. It shows the response of a quantity demanded for goods or services to a change in price. PED is almost always negative, i.e. an increase in price will cause a reduction in demand.

Price elasticity of supply (PES) is the ratio of percent change in a quantity supplied, S, to the percent change in price, P. It measures the sensitivity of the quantity of goods and services to the change in market price for those goods or services. PES is almost always positive, i.e. an increase in price will cause an increase in supply. A value of PES less than one is considered inelastic.

23

Methods for Total Plant and Total Product Cost Estimation

Introduction: To evaluate the profitability of a proposed new plant, the plant and total product costs are estimated. These cost estimates are done in stages and are used with revenue forecasts to obtain an estimate of the profitability. Plant costs includes equipment, installation and set-up costs as well as other costs related to purchase of land etc., and they are summarized in the list in Table 2. Also, total product costs include manufacturing costs and general expenses, and they are summarized in the list in Table 3.

Economic and process models are required for profitability evaluations at several levels of detail and complexity, depending on the point in the plant design. There are five stages in estimating the proposed plant's profitability that are classified in the list below (Garrett, 1989, Turton, et al., 1998). Cost estimation for the plant increases in complexity and cost with these steps.

1. Order of Magnitude: A preliminary design based on estimations of overall plant performance and cost of production derived from existing plants. It is based on preliminary material balances and block process diagrams. Accuracy is $\pm$ 40-100%, and cost is $5,000- $20,000.

2. Study Estimate, Preliminary, or Major Equipment: Design sizing of major process units and estimating associated costs using vendor quotes and estimation methods. It is based on a process flow diagram (PFD) with complete material and energy balances and major equipment sized. Equipment and plant costs are evaluated using estimating factors. Accuracy is $\pm$ 20-50%, and cost is $20,000- $50,000.

3. Preliminary, Budget Authorization, Scope Estimate: Performed by professional estimators, e.g. company's engineering design department or a contractor using single vendor quotes on major equipment. It is based on the PFD from step 2 and approximate equipment layout, estimates for piping, instrumentation and utilities are made. Accuracy is $\pm$ 15-25%, and cost is $50,000- $200,000.

4. Definitive, Project Control Estimate: Detailed design to prepare final process flow diagram (PFD), preliminary piping and instrumentation diagram (P&ID), specification for equipment using competitive bids, utilities, and instrumentation for possible approval to construct the plant. Accuracy is $\pm$ 5-15%, and cost is $200,000- $800,000

5. Detailed Estimate, Firm or Contractor's Estimate: Competitive vendor quotes, complete drawings and specifications, final PFD, P&ID, piping isometrics, etc. Accuracy is $\pm$ 5-10%, and cost is 1-5% of total plant cost.

Table 2 Partial List of Components in a Plant Cost Estimate, after Garrett, 1989.

On-Site Facilities

Process Equipment, such as:

Towers, columns	Dryers	Pumps	Filters, centrifuges
Heat exchangers	Reactors	Tanks	Evaporators

Installation Costs (Labor to install equipment) and including:

Insulation	Piping	Utilities	Electrical
Instruments	Yard improvements		Foundations
Painting	Railing, catwalks		Platforms
Buildings	Inventory, supplies, catalyst		Safety Equipment
Fireproofing	Product storage, handling		Environmental facilities

  Construction Expense:

Engineering,	Changes, additions	Licenses, fees
Tools,	Temporary facilities	Supervision, overhead
Construction equipment	Accounting, scheduling, planning	
Models: computers, software	Environmental, safety, studies, reports, etc.	

Company Costs:

Design, drafting	Engineering	Licensing fees	Research and development
Owner's inspection	Feasibility studies		Market research

Off-Site Facilities

Typical Utilities

Boilers, condensate and makeup water systems	Generators (including cogeneration)
Standby generators or battery assemblies	Main power transformer stations
Fuel storage and distribution facilities (oil, coal, gas, etc.)	
Plant-wide air conditioning facilities	Sewage collection (and treatment)
Plant-wide paging, emergency communication system	
Inert gas systems, fire fighting equipment and systems	Flares, stacks, waste gas treatment
Compressed, instrument air, cooling towers	Distribution systems, refrigeration

Service Buildings and Related Facilities

Employee lockers, showers, time-card, lunch, restrooms

Eng., technical service facilities

Office building (management, sales, accounting, admn.)

Laboratory, R & D, environmental

Shipping, receiving office, supply warehouse   Maintenance buildings, shops

Inventory (raw materials, products, supplies) storage

Product Sales

Loading, unloading rail spurs, docks, forklifts, loaders, etc. Packaging facilities

Shipping equipment (trucks, railcars, ships, barges, etc.)   Warehouses

Distant storage, reshipping facilities

Environment

Air, water, and ground monitoring equipment and control

Water treating and reuse facilities   Solid or liquid waste shipping facilities

Solid or liquid waste processing, handling equipment   Incineration equipment

Start-up, Working Cap

Table 3. List of Components in Total Product Cost Estimate, after Peters and Timmerhaus, 1991

Component	Power and Utilities	Production/Cost group	Manufacturing/General	Total
Raw materials		Direct Production Costs	Manufacturing Costs (C_M)	Total Product Cost (C_T)
Operating labor				
Operating supervision				
Steam	Power and Utilities			
Electricity				
Fuel				
Refrigeration				
Water				
Maintenance and repairs				
Operating supplies				
Laboratory charges				
Royalties				
Catalysts and solvents				
Depreciation		Fixed Costs		
Taxes on property				
Insurance				
Rent				
Royalties, Interest		Plant Overhead Cost		
Indirect labor charges				
Fringe benefits				
Medical				
Safety and protection				
General plant overhead				
Payroll overhead				
Packaging				
Restaurant				
Recreation				
Salvage				
Control laboratories				
Plant superintendent				
Storage facilities				
Executive salaries		Administrative Expenses	General Expenses (C_G)	
Clerical wages				
Engineering and legal costs				
Office maintenance				
Communications				
Sales offices		Distribution and Marketing Expenses		
Salesmen expenses				
Shipping				
Advertising				
Technical sales service				
Research and development				
Gross-earnings expense				

Estimating Equipment Cost for the Plant: In order to calculate initial plant cost estimates, it is essential to know or have a general idea of the individual equipment costs. These can be obtained from either the manufacturer's price quotations, capital cost estimation programs or through estimating charts such as those given in Appendix A. These charts give approximate values and provide valuable information when the actual prices cannot be obtained. The following information is required to estimate the cost of the equipment for the proposed plant.

1. Prepare a flow sheet for the process or operation to be estimated which includes all of the major equipment and the basic auxiliary or general plant facilities that are directly involved. A flowsheeting program is recommended.

2. Complete the material and energy balances around each piece of equipment to size the equipment. A flowsheeting program is recommended.

3. Size the equipment with the precision required to obtain the dimensions and conditions needed for vendor's quotes.

4. Analyze the process to determine the cost factors that are appropriate to use to estimate on-site and off-site costs as shown in Table 2.

      Manufacturers' Quotations: With the above information available, vendors can be contacted for estimates of the direct purchase price, and shipping and installation charges can be approximated. This is the preferable method of obtaining equipment cost estimates, but it may not be appropriate at certain times when proprietary information could be released. *Chemical Week*, *Chemical Engineering*, *Chemical Processing*, *Hydrocarbon Processing* and *American Laboratory* give extensive lists of vendors in publications. However, obtaining price estimates may be lengthy if special design details and parameters are required, as described by Garrett, 1989.

      Capital Cost Estimation Programs: The Aspen Capital Cost Estimator program from Aspen Technology is based on Aspen Icarus technology and estimates plant capital equipment costs with associated plant bulks using design-based installation models. Stored in Icarus systems are design and cost models for over 250 kinds of liquid, gas and solids handling and processing equipment, more than 60 kinds of plant bulk items, approximately 70 kinds of site preparation work, and nearly a dozen types of buildings. Installation bulk models are used to develop installation quantities and field manpower and costs to install equipment and plant bulks. To support these design and cost models, Icarus systems contain design procedures and costs data for hundreds of types of materials of construction for general process equipment, vessel shells and internals, tubing, castings, linings, packings, clad plates, piping, steel and electrical bulks, Aspen Technology, 2013.

      Five other capital cost estimation programs were described and compared by Feng and Rangaiah, 2011, which include CAPCOST, EconExpert, AspenTech Process Economic Analyzer (PEA), Detailed Factor Method (DFP) and Capital Cost Estimation Program. They reported that the programs are based on different methods and developed in different platforms; all five are user friendly and useful for estimating the capital cost of chemical plants. Aspen PEA had the most equipment types available and DFP had the least. The plant costs for seven cases differed, as much as 81% and individual equipment capital costs for different programs may not be comparable. Total module costs obtained by Aspen PEA were reported to be more than 60% higher than those by CAPCOST that was mainly due to the high costs of towers and vessels in Aspen PEA. A university developed stand-alone program, CAPCOST_2011.xls, comes with the text by Turton, et al., 2012 that is uses equipment modules and is programmed in Microsoft Excel.

Estimating Charts: Cost estimates can be made rather quickly with charts that are available from published sources and company past experience. Estimating charts are available from a number of sources including Hall, et.al, 1984, Douglas, 1988, Peters and Timmerhaus, 1991, and Garrett, 1989. The ones by Garrett are very complete, and he was careful to adjust the cost to 1987. A number of the charts in Garrett have been reproduced in Appendix A along with some of his charts for estimating complete plant costs and manufacturing costs. Garrett states that actual prices can range from +100% to -50% depending on quality, designs and specifications, but the results obtained using these charts are in the generally correct range and represent good to high quality equipment.

Estimating charts are log-log plots, and the cost data is approximately a straight line. These charts are based on the following scaling equation. The cost, C, can be predicted over a range of equipment size (capacity) values, S, if one value of C_1 and S_1 is known along with the exponent, n.

$$C = C_1 (S/S_1)^n \qquad\qquad (1)$$

where C_1 and S_1 are a known cost and equipment size (capacity). This equation has lead to a "0.6 rule" that n is approximately 0.6 as a rule of thumb. This exponent is given on the estimating charts in the Appendix.

In Perry's, 1997 Table 25-49, an extensive listing of equipment is given with sizes, costs and exponents, so the above equation can be used to estimate purchased cost. This equation is used to include the effects on temperature, pressure, materials of construction (Table 4), inflation (Figure 8 and 9), and location in the world (Figure 10) as described below. McConnell, 2009, gives additional data for international locations based on 2009 data.

Table 4 Typical Factors for Converting Carbon Steel Costs to Equivalent Alloy Costs, after Table 9-58, Perry's, 1997

Material	Pumps, etc.	Other Equipment
All carbon steel	1.00	1.00
Stainless steel, Type 410	1.43	2.00
Stainless steel, Type 304	1.70	2.80
Stainless steel, Type 316	1.80	2.90
Stainless steel, Type 310	2.00	3.33
Rubber-lined steel	1.43	1.25
Bronze	1.54	-
Monel	3.33	-

Material	Heat Exchangers
Carbon steel shell and tubes	1.00
Carbon steel shell, aluminum tubes	1.25
Carbon steel shell, monel tubes	2.08
Carbon steel shell and tubes	1.00
Carbon steel shell, 304 stainless tubes	1.67
304 stainless steel shell and tubes	2.86

Adjusting Cost Estimates for Inflation: Estimating charts are based on a point in time when purchase prices were obtained. These prices need to be corrected for inflation to the present time, and there are several indices that are available to do this correction. Some are shown in Figure 8. The one used most frequently is the Chemical Engineering Plant Cost Index (CEPCI) that been available for 40 years, and Vatavuk, 2002, describes an update for annual plant cost indices and process equipment indices. Other cost indices are the Nelson-Farrar Refinery Construction Index that is published in the *Oil& Gas Journal* with a number of categories related to refining and the Engineering News-Record Constructions Index that is published in the *Engineering News – Record* with materials and labor categories.

To use the index (Equation 1), the cost obtained from an estimating chart C_1 would be multiplied by the ratio of the current index I_2 (381.5 - Oct., '98) to the index for the chart I_1 which is 320 for the ones in Appendix A (1987), i.e., $C_2 = C_1 (I_2 / I_1)$.

CEPCI is available online at http://www.che.com/pci/ for access to the CEPCI database, including annual archives (1947 to present) and monthly data archives (1970 to present). It is published in each issue of *Chemical Engineering* magazine

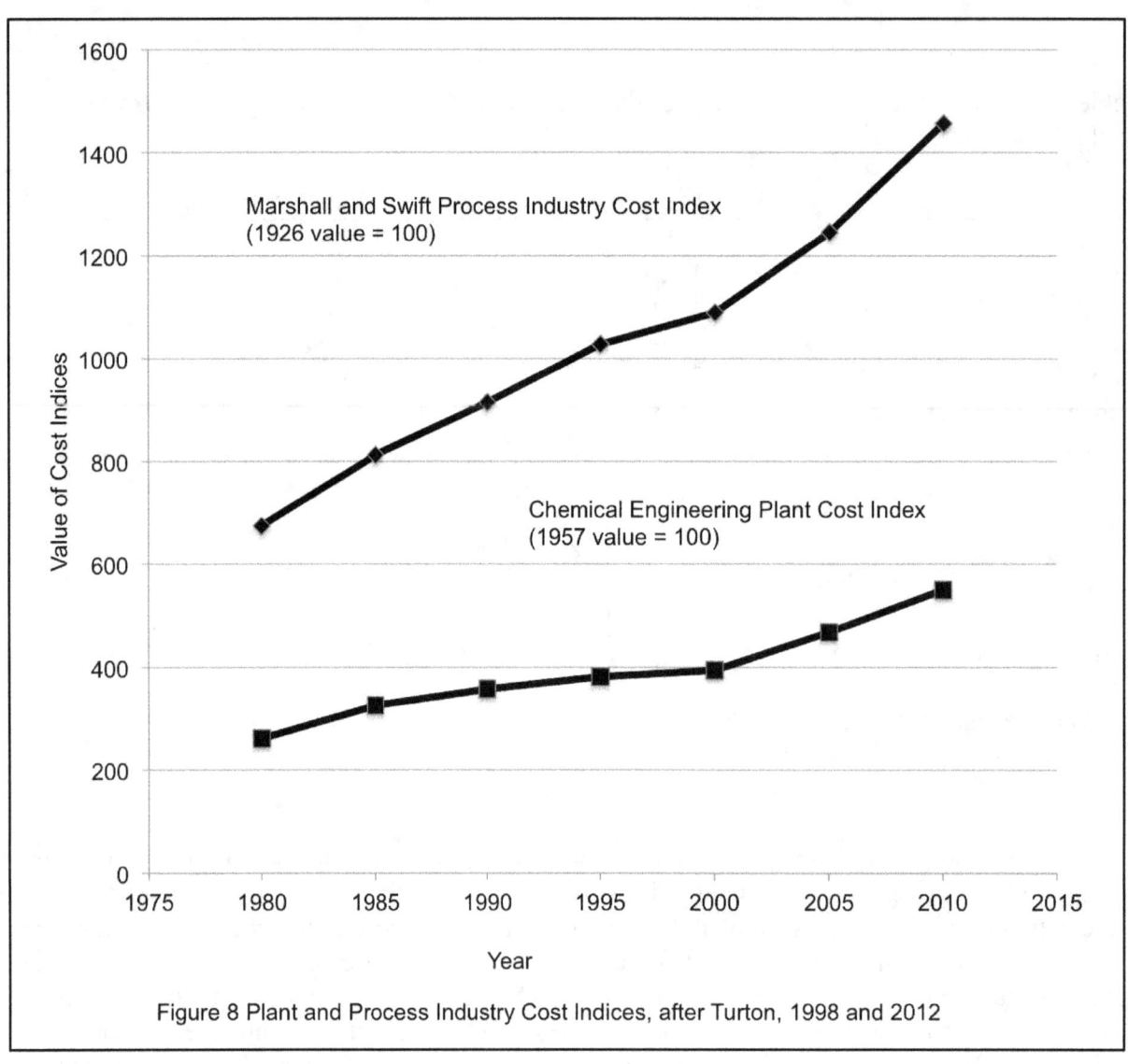

Figure 8 Plant and Process Industry Cost Indices, after Turton, 1998 and 2012

Chemical Engineering Plant Cost Index (CEPCI) 1957-59 =100

	May, '11 Prelim.	April '11 Final	May '10 Final
CE Index	551.7	582.3	558.2
Equipment	707.3	708.0	670.2
Heat exchangers and tanks	673.0	671.4	629.9
Process machinery	663.7	665.3	631.8
Pipes, valves, fittings	861.8	867.9	828.3
Process instrumentation	440.1	441.7	424.8
Pumps and compressors	904.4	904.7	903.1
Electrical equipment	503.0	502.6	473.2
Structural support and misc.	756.7	762.8	697.5
Construction labor	326.0	325.8	327.8
Buildings	515.2	517.1	513.9
Engineering and supervision	332.9	333.6	339.3

Annual Index

1991	361.3
1992	358.2
1993	359.2
1994	368.1
1995	381.1
1996	381.7
1997	386.5

Figure 9 Chemical Engineering Plant Cost Index, Source: Chemical Engineering Magazine, 2011

A comparison of cost indices normalized to 2006 as the value '1'are shown in Figure 11 from McConnell (Worley-Parsons), 2009 for the years 2004 to 2009. During 2007 and 2009, a steep increase in the cost indices was observed, and these costs were attributed to the steep rise in equipment costs tied to a spike in the price of metals and demand for equipment during this time. During the initial part of 2009, corrections of these costs to a more moderate escalation were observed related to the slowdown in the global economy. Sources for information included the Chemical Engineering Magazine and the U. S. Corps of Army Engineers (2009). The Army Corps of Engineers and Marshall and Swift indexes are for power plants and are combined equipment and labor indices. The Chemical Engineering index shown here is for equipment.

Australia	1.3	Malaysia	0.8
Austria	1.0	Middle East	1.1
Belgium	1.0	Newfoundland	1.2
Canada	1.15	New Zealand	1.3
Central Africa	~2.0	Norway	1.1
Central America	1.0	South America (North)	1.35
Denmark	1.0	South America (South)	2.25
Eire	0.8	Sweden	1.1
Finland	1.2	Switzerland	1.1
France	0.95	Turkey	1.0
Germany	1.0	U.K.	0.9
Greece	0.9	U.S.	1.0
Holland	1.0	Yugoslavia	0.9
Italy	0.9		

China
(imported element)   1.1
(indigenous element)   0.55

North Africa
(imported element)   1.1
(indigenous element)   0.75

India
 (imported element)   1.8
(indigenous element)   0.65

Spain(imported element)   1.2
(indigenous element)   0.75

Note: Increase a factor by 10% for each 1,000 miles, or part, for plants located from a major industrial area.

Figure 10 Foreign Location Factors, after Garrett, 1989

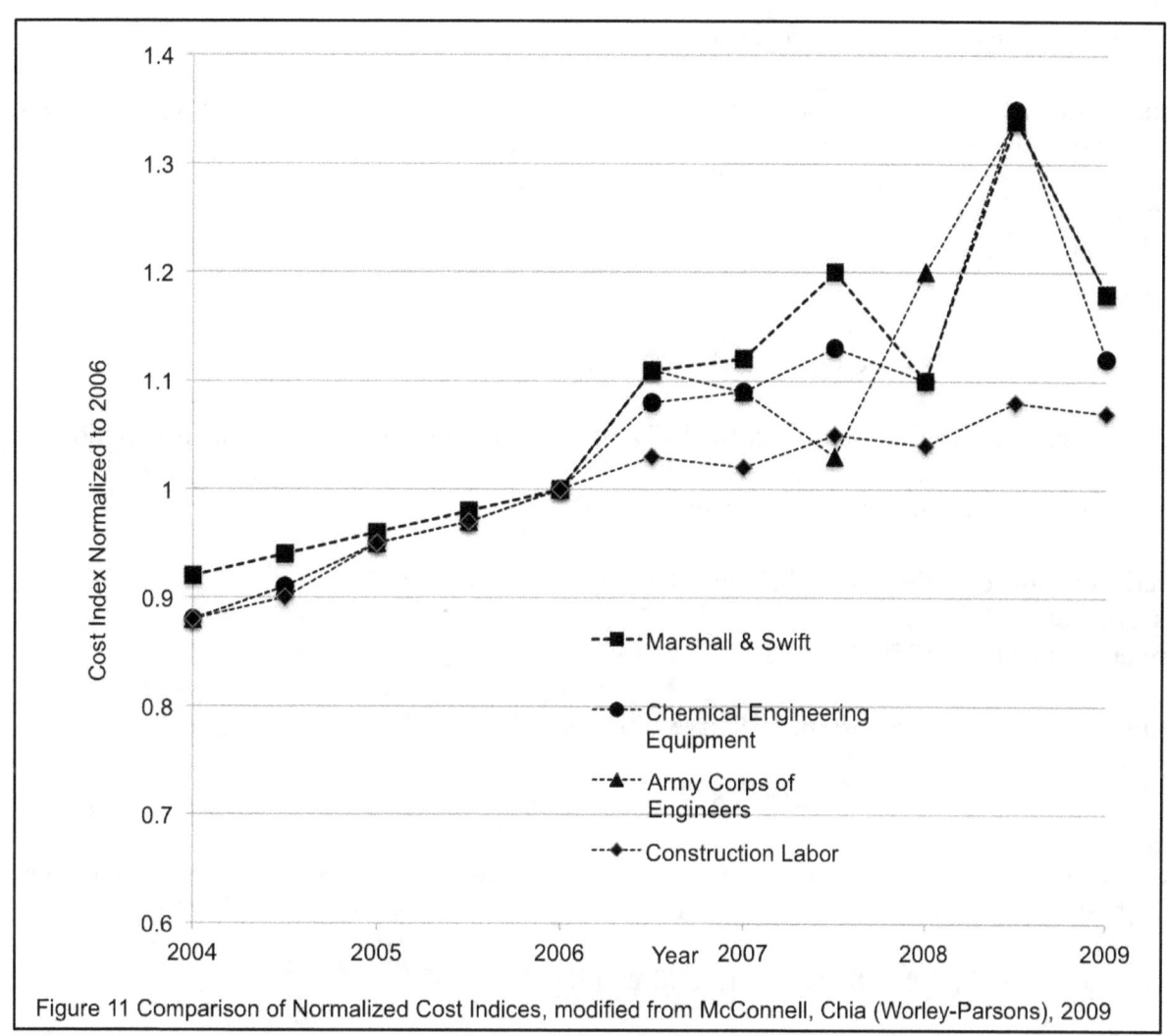

Figure 11 Comparison of Normalized Cost Indices, modified from McConnell, Chia (Worley-Parsons), 2009

Installed Cost Estimation: The cost read from the charts in Appendix A is the purchase price of the equipment, and the charts contain two other multipliers used to estimate the installed and module costs. These installation and module factors are used to estimate the total cost to purchase the equipment, have it shipped to the plant, installed and be ready to operate. The installed cost, Garrett, 1989, is used when replacement equipment is being installed in an existing plant. The module cost is larger, and it is used for a new plant or plant expansion. Comparable information is given in Perry's, 1997, also.

Example 2. Estimation of Installed Cost for a Heat Exchanger

Estimate the installed cost of a floating-head shell-and-tube heat exchanger that was purchased in 1992. The exchanger has a carbon steel shell and stainless steel tubes. It has a total heat transfer area of 2,200 square feet and is rated for 250 psig and 350 °F. The equipment cost, based on 1979 prices, is given by:

$$C = 117 \, A^{0.65}$$

where A is the area in square feet, and C is the cost in 1979 dollars.

The Chemical Engineering cost index for 1979 was 230. In 1992 it was 360. The installation factor for heat exchanger is 3.29. (See Figure A-6.) Special factors, relative to a "standard" low-pressure, low-temperature, all carbon steel exchanger are:

Materials factor for carbon steel shell and stainless steel tubes = 2.81
Pressure factor for 250 psig = 1.20
Temperature factor for 350 °F = 1.00

Calculating the purchased cost in 1979 gives:

$$C_{purchased, \, 1979} = 117 \, (2,200)^{0.65} = \$17,408$$

The installed cost in 1992 is the purchased cost multiplied by the various factors for materials, 2.81, pressure, 1.20, temperature, 1.00, installation, 3.29, and inflation, (360/230) as follows:

$$C_{installed, 1992} \quad = \$17,408(2.81)(1.20)(1.00)(3.29)(360/230) = \$302,300$$

Plant Cost Estimation: Having obtained the purchased equipment installed cost, the other components of the plant cost can be estimated using the estimating factors given in Table 4, from Garrett, 1989. These costs include the cost of land and foundations or structural members to support the equipment. Then there are costs related to piping, conveyer lines, electrical lines, switchgear, instruments etc. Further, there are costs of buildings for labs, offices, warehouses and maintenance. Also, there are costs for off-site facilities such as roads, utility services and transportation equipment.

Table 5 Plant Cost Estimating Factors, after Garrett, 1989

Component	Plant Cost Factor, Fraction of Total Purchased Equipment Cost
Purchased equipment	1.00
Piping	0.15-0.70
Electrical	0.10-0.15
Instrumentation	0.10-0.35
Utilities	0.30-0.75
Foundations	0.07-0.12
Insulation	0.02-0.08
Painting, fireproofing, safety	0.02-0.10
Yard improvements	0.05-0.15
Environmental	0.10-0.30
Buildings	0.05-1.00
Land	0.00-0.10
Construction, engineering	0.30-0.75
Contractors fee	0.10-0.45
Contingency	0.15-0.80
Total	2.51-6.80

Usual limits for total factors

Minimum (solids processing)	3.00
Average (mixed processing)	4.00
Maximum (fluid processing)	5.00

Other capital requirements	*Additional Factor, Percent of Total Plant Cost*
Off-site facilities	0-30
Plant start-up	5-10
Working capital	10-20 or 10-35 % of mfg. cost

In Table 5, the estimating factors are given as a fraction of the purchased equipment cost for the plant. These can be evaluated, and summed to obtain an estimate of the total plant cost. According to Table 5, the total plant cost would be between 2.51-6.80 times the purchased equipment costs. Additional costs as a percentage of the total plant costs are added for off-site facilities, start-up and working capital. Douglas, 1988 recommends using 2.36 times the onsite costs to calculate the total plant cost. The onsite costs are the installed equipment costs for items on the process flowsheet. Also, the Lang factor can be used, as described by Turton, 1998 where the purchased equipment cost are multiplied by 4.74 for a fluid processing plant, by 3.63 for a solids-fluids processing plant and

There are estimating charts for the average cost of complete plants, and Garrett, 1989 presents a number of these charts for a wide range of plants. One of these is given in Appendix A for alkylate detergents, aniline, allyl chloride and acetic anhydride. However, he points out their limitations including age, lack of environmental control costs, and relatively small size range. Also, Perry's 1997 has Table 9-48 which gives capital cost data for processing plants.

The costs presented in the estimating charts are for average U.S. prices and locations. Garrett, 1989 presents foreign location factors to adjust plant cost estimates for other parts of the world, and these are shown in Figure 10. As with other estimations, he warns that these factors are an oversimplification of the problem, give approximate values, but help provide a more realistic estimate of plant costs in other countries. Perry's 1997 provides similar information in Table 9-55 for 1993 values.

Example 3 Study Estimate for the Aniline Plant Design

Process Description

The process flow diagram (PDF) for a 100 million pound per day plant for aniline from the reaction of ammonia and phenol is given in Figure 12 from an Excel spreadsheet. This will be used for a study estimate using the major process equipment given in the PFD. In this process, phenol and aniline are reacted with a catalyst to produce aniline and water with a 98% conversion. The effluent from the reactor is sent to product purification facilities that include three distillation columns. The first column removes ammonia from the mixture that is recycled. The second column separates water and the remaining ammonia to dry the product. The third column removes the byproduct diphenylaniline. A preliminary material balance is shown on the PFD along with information associated with an estimate for the size of the equipment and utilities required.

The installed costs of the equipment on the PFD were determined using the charts given in Appendix A and the program CAPCOST for comparison. These results are given in Figure 13 from an Excel spreadsheet, including numbers that are not significant. These charts are based on 1987 costs. The 1987 Chemical Engineering Plant Cost Index (CEPCI) is 330. The 2000 value is 394 (Vatavuk, 2002). In this figure there is a comparison with the program CAPCOST from Turton, et al., 1998, (CEPCI of 382, 1996). These results are within the accuracy of the estimating procedures.

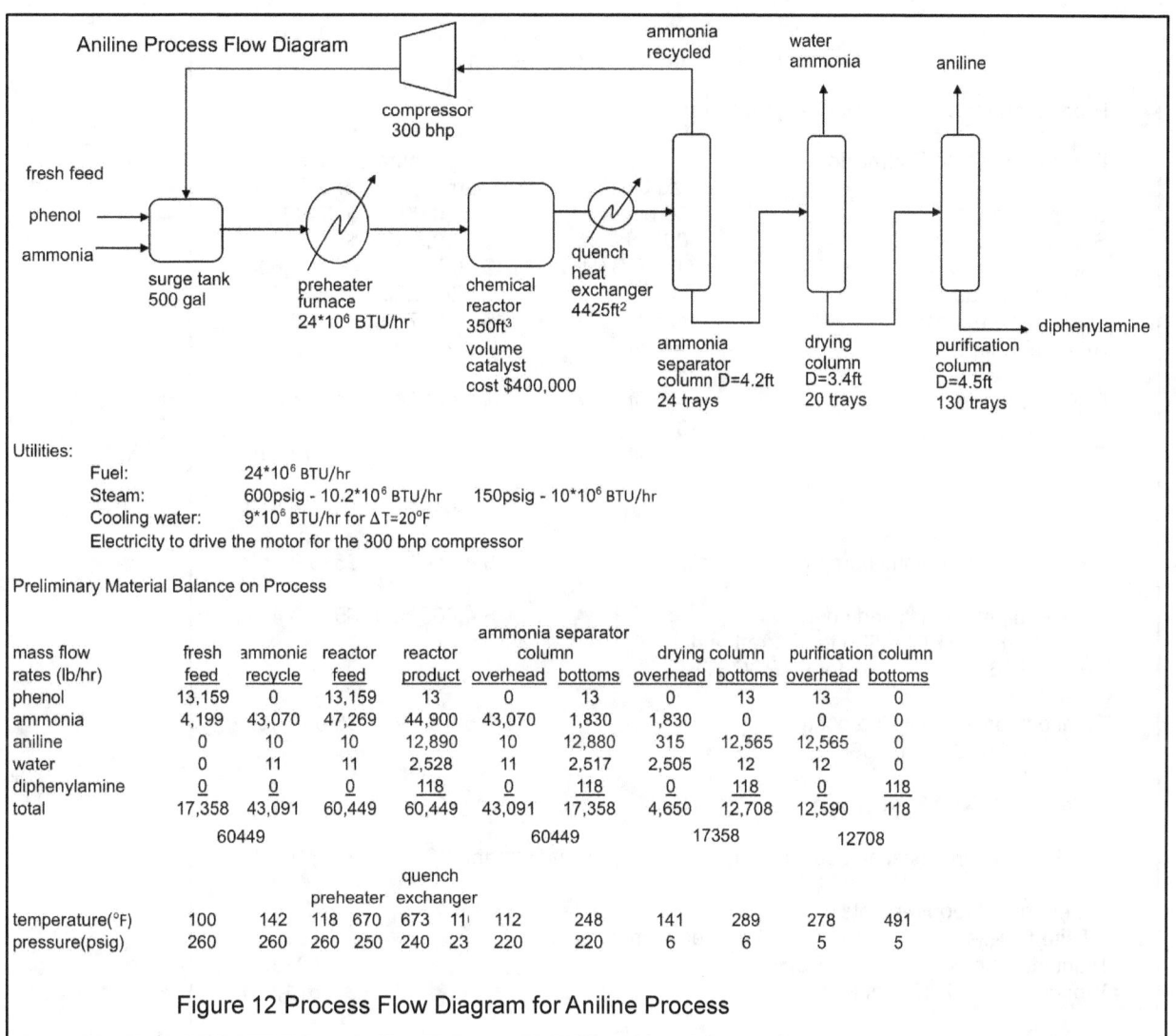

Aniline Process Flow Diagram

ammonia recycled

water ammonia

aniline

compressor
300 bhp

fresh feed

phenol

ammonia

surge tank
500 gal

preheater
furnace
24*10⁶ BTU/hr

chemical
reactor
350ft³
volume
catalyst
cost $400,000

quench
heat
exchanger
4425ft²

ammonia
separator
column D=4.2ft
24 trays

drying
column
D=3.4ft
20 trays

purification
column
D=4.5ft
130 trays

diphenylamine

Utilities:

Fuel:	24×10^6 BTU/hr
Steam:	600psig - 10.2×10^6 BTU/hr 150psig - 10×10^6 BTU/hr
Cooling water:	9×10^6 BTU/hr for $\Delta T = 20°F$
	Electricity to drive the motor for the 300 bhp compressor

Preliminary Material Balance on Process

mass flow rates (lb/hr)	fresh feed	ammonia recycle	reactor feed	reactor product	ammonia separator column overhead	ammonia separator column bottoms	drying column overhead	drying column bottoms	purification column overhead	purification column bottoms
phenol	13,159	0	13,159	13	0	13	0	13	13	0
ammonia	4,199	43,070	47,269	44,900	43,070	1,830	1,830	0	0	0
aniline	0	10	10	12,890	10	12,880	315	12,565	12,565	0
water	0	11	11	2,528	11	2,517	2,505	12	12	0
diphenylamine	0	0	0	118	0	118	0	118	0	118
total	17,358	43,091	60,449	60,449	43,091	17,358	4,650	12,708	12,590	118

60449   60449   17358   12708

	fresh feed	ammonia recycle	preheater	quench exchanger	ammonia separator overhead	ammonia separator bottoms	drying overhead	drying bottoms	purification overhead	purification bottoms
temperature(°F)	100	142	118 670	673 11	112	248	141	289	278	491
pressure(psig)	260	260	260 250	240 23	220	220	6	6	5	5

Figure 12 Process Flow Diagram for Aniline Process

Aniline Plant Capital Cost Estimation

Production rate = 100 million pounds per year of aniline

Equipment Cost Estimation

	Size Units	Installed Cost Garrett	CAPCOST
Compressor	300 bhp	$ 650,000	$ 549,954
Surge Tank	500 gal	$ 10,500	$ 28,672
Preheater Furnace	$24*10^6$ BTU/hr	$1,071,000	$ 866,688
Chemical Reactor	$350ft^3$	$1,288,260	$1,505,030
Quench exchanger	$4425ft^2$	$ 35,712	$ 98,236
Ammonia Separator	D=4.2ft	$ 238,945	$ 388,113
trays	24	-	-
Drying Column	D=3.4ft	$ 171,666	$184,410
trays	20	-	-
Purification Column	D=4.5ft	$ 958,035	$1,518,416
trays	130	-	-
	Subtotal	$4,424,118	$5,139,519

10% for miscellaneous pumps and vessels	$ 442,412	$513,952
Total equipment installed cost (1987 CEPCI = 330 for Garrett and 1996 CEPCI = 382 for CAPCOST)	$ 4,866,530	$5,653,471
Total equipment installed cost (2000, CEPCI = 394)	$ 5,810,342	$5,831,067

Total Plant Cost Estimation

Total equipment installed cost on-site or fixed capital investment, FCI	$ 5,810,342

Other capital requirements

Offsite facilities, 25% of on-site installed equipment cost	$ 1,452,585
Plant start-up costs, 10% of FCI	$ 581,034
Working Capital, 15% of FCI	$ 871,551
Total plant cost or total capital investment, TCI	$ 8,715,512

Figure 13 Equipment and Total Plant Cost Estimation

Total Product Cost Estimation: For plant profitability analysis, the annual cost of manufacturing the product is required. The main components of the manufacturing costs are those for raw materials, utilities and labor as shown in Table 6. These major components have to be itemized from the process flowsheet and material and energy balances. The other costs can be estimated as a percentage of labor, plant costs and sales as shown in the table.

One of the best sources of data is information from similar plants. If these records are available, then quick and reliable estimates of manufacturing costs and general expenses can be made. Alternatively, estimates can be made based on manning charts, local wage scales, manufacturer's expected maintenance schedules or by using the factoring method.

In addition to manufacturing costs, there are capital related costs and general expenses or sales related costs as shown in Tables 3 and 5. These include expenses related to administration, distribution, marketing, and research and development. These costs are relatively constant and do not vary with the production rate.

Raw materials are major items of costs, and raw materials that are calculated from process material balances. The cost of raw materials should be based on the amount of raw materials actually consumed.   Certain materials such as catalysts may be recoverable. The initial cost of such recoverable materials (e.g., catalysts, filters) should be included in the plant cost. The cost of regeneration, catalyst makeup and replacement should be included in manufacturing cost. In addition to the cost of raw materials calculated from material balances, there is usually a small allowance for extra materials because of the plant's inevitable losses.

For initial cost estimation, market prices can be used for calculating raw materials cost. These values are regularly published in various periodicals and journals such as the *Chemical Marketing Reporter*, *Chemical Buyers Directory*, *Chemical Cyclopedia*, and *European Chemical News*. Also, information is available at the web sites www.chemexp.com and www.chemweek.com. Actual prices may differ from the published prices, since they are usually negotiated depending on factors such as quality, quantity, duration of contract, and state of the market. A better source for such information would be direct price quotation from prospective suppliers. The cost of transportation is usually not included in such price quotations. Hence, freight, transportation, and loading/unloading costs should be included in the raw materials cost. Some typical values are given in Table 7 from Brown, 2000.

Table 6 Total Product Cost Estimating Factors, after Garrett, 1989

Manufacturing Costs

Raw materials	Itemize
Utilities	Itemize
Operating labor	Itemize
Interest (on loans, if any)	Itemize

Labor related costs

Payroll overhead	22 - 45 % of labor
Supervisory, miscellaneous labor	10 - 30 % of labor
Laboratory charges	10 - 20 % of labor
Total	42 - 95 % of labor
(Typical total)	60 % of labor

Capital related cost

Maintenance	2 - 10 % of plant cost
Operating supplies	0.5- 3 % of plant cost
Environmental	0.5- 5 % of plant cost
Depreciation	5 - 10 % of plant cost
Local taxes, insurance	3 - 5 % of plant cost
Plant overhead costs	1 - 5 % of plant cost
Total	12 - 38 % of plant cost
(Typical total)	26 % of plant costs

General Expenses or Sales Related Costs

Patents and royalties	0 - 5 % of sales
Packaging, storage	0 - 7 % of sales
Administrative costs	2 - 10 % of sales
Distribution and sales	2 - 10 % of sales
R & D	0.5 - 4 % of sales
Total	4.5 - 36 % of sales
(Typical total)	20 % of sales

Table 7 Shipping Costs, after Brown, 2000

Shipping Method	Capacity	Price($/mile)
Rail Car	20,000 gal	2.05
Tank Truck	45,000 lb	2.75
Semi-Trailer, dry van	45,000 lb	1.35

Operating labor cost is another element of manufacturing cost that is itemized. Labor required can be estimated from either company experience, or graphs like Figure 14. Once the process flow diagram is available, it can be analyzed to estimate the amount of labor required. The simplest way is to estimate the labor requirements for each major piece of equipment. Usually, each major piece of equipment and its supporting facilities and controls requires one operator, according to Garrett, 1989. This can be compared to operating labor requirements that are related to the plant capacity in the graphs in Figure 14. Also, occupational employment and wages for chemical plant and systems operators is available from the Bureau of Labor Statistics of the U. S. Department of Labor.

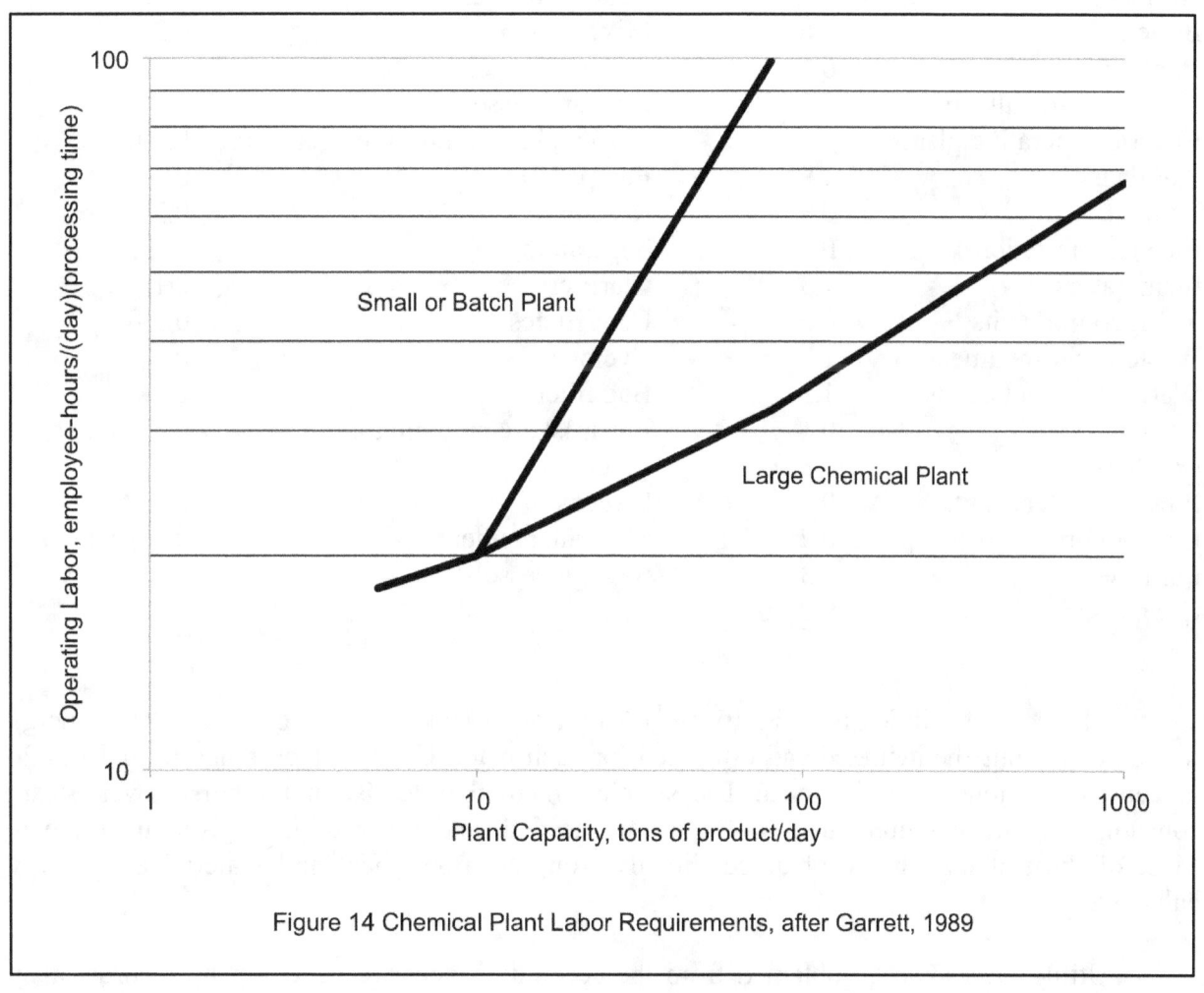

Figure 14 Chemical Plant Labor Requirements, after Garrett, 1989

In Table 8 results are given by Brown, 2000 to estimate direct labor by equipment being operated. This table has been updated to 1998 from Ulrich in 1982. Also, Perry's 1997 includes an equation that can be used to compute operating labor estimates similar to the results given in Table 8.

Table 8 Estimating Direct Labor Costs, from Brown, 2000

Equipment	No. Operators/unit/shift	Equipment	No. Operators/unit/shift
Auxiliary facilities		Gas-solid contacting equipment	0.05-2.0
Air plants	0.6	Heat exchangers	0.05
Boilers	0.6	Mixers	0.2
Cooling towers	0.6	Process Vessels	0
Water demineralizers	0.3	Pressure vessels	
Electric generating plants		Towers, inc. pumps & exchangers	0.1-0.3
Stationary	1.9	Pumps	0
Portable	0.3	Reactors	0.3
Electric substations	0	Separators	
Incinerators	1.3	Clarifiers	0.1
Refrigeration units	0.3	Centrifuges	0.3-0.1
Waste water treatment	1.3	Cyclones	0
Water treatment plants	1.3	Bag filters	0.1
Conveyers'	0.1	Electrostatic precipitators	0.1
Crushers	0.3-0.6	Filters	0.6
Power recovery units	0	Screens	0.03
Evaporators	0.2	Size-enlargement	0.05-0.2
Furnaces	0.3	Storage vessels	0
Compressors	0.05-0.1		

The cost of utilities is a significant fraction of the total operating cost, and the energy requirements must be itemized and estimated for each plant. Usually, they range from 10 to 20 percent of the total operation cost. The simplest method is to list motor horsepower, steam consumption, fuel requirements, cooling water needed, and other utilities given in Table 9. Most of these items can be obtained directly from the flow sheet and material and energy balances.

Utility prices can be obtained from the accounting department, company records or by inquiry to outside sources. Some typical utility costs are given in Table 9. There are no satisfactory shortcut methods to estimate utility requirements, according to Perry's, 1997.

Table 9 Typical Utility Costs, after Garrett, 1989 (Southern California, CE Index 320)

Electricity	$0.08 /kW hr
Gas	$7.00/ MM Btu
Fuel Oil, low sulfur	$7.00/ MM Btu ($20/bbl)
Steam, 250 psig	$12.00/MM Btu
Cooling tower water	$0.05 M gal
Process Water: city water	$0.20 M gal
well water	$0.10 M gal
Recycled process water	$0.25/10^3 gal
Recycled cooling tower water	$0.20/10^3 gal
Demineralized water	$5.00/10^3 gal
Instrument air (dry)	$0.30 /M scf
Inert gas, low pressure	$1.00 /M scf
Nitrogen, purchased	$2.20 /10^3 scf
Refrigeration (ammonia to 30°F)	$1.50/ ton-day (288,000 Btu removed)

In summary, raw materials, labor and utility requirements are itemized in a detailed cost analysis, whereas, costs of other elements such as maintenance, repairs, patents and royalties, packaging, and laboratory charges are taken to be a percentage of either labor, capital, or sales related costs. The estimating factors for such elements are given in Table 5, after Garrett, 1989.

Capital related costs can be estimated using the estimating factors given in Table 5 as a fraction of the plant cost. Unlike the direct production costs, fixed charges do not depend on production. They represent those costs that continue whether the plant is in production or not. They are related to capital investment and include expenses for depreciation, local property taxes, financing, and insurance charges. Insurance and property taxes depend to a large extent on the location of the plant, hazards involved, amount of facilities provided and upon local regulations.

Equipment undergoes wear and tear each year, and its value decreases. Consequently, it can be depreciated as a business expense based on its economic life. There are different ways for estimating the depreciation. Some provide for higher depreciation charges during the early years of the project and lower charges in subsequent years. Also, the economic life for machinery and equipment is shorter than the economic life of buildings. Methods for computing depreciation are described later in the chapter. For initial estimates, straight-line depreciation is used typically. Straight-line depreciation is computed by taking the original plant cost less the salvage value and dividing by the economic life of the plant. This gives an annual cost that is a business expense.

Example 4 Total Product Cost Estimation for Aniline Process

An estimate of the total product cost is given in Figure 15 from an Excel spreadsheet. This estimate uses the results given in the tables in this section

Manufacturing Costs

Utilities

		Mbtu/hr	cost $/1,000lb	Btu/lb	Enthalpy $/yr
steam	600psig	10.2	4.52	728	$538,429
	150 psig	9	3.4	858	$303,218
electricity	300bhp compressor		$0.04/kwhr 70% eff.		$108,728
fuel	24 million Btu/hr		$4.0/Mbtu		$816,192
cooling water	9.0 million Btu/hr		$0.03/1000 gal		$13,826

Total utility cost		$1,780,393

Raw Materials

	$/lb	lb/hr	$/yr
Phenol	0.15	13159	$16,781,673
Ammonia	0.075	4199	$2,677,492
Catalyst (negligible)			-

Total raw materials	$19,459,165

Operating Labor

Operating labor,$30/hr-4.5 op./shift pos.4 shift pos $4,591,080

Labor Related Costs
Superv. & support, 35% Op. Lab. $1,606,878

Capital Related Costs

Maintenance & supplies, 5% FCI	$198,953
Plant Overhead, 70% Op. Lab. + 3% FCI	$3,333,128
Property taxes & ins., 3% FCI	$119,372

General Expenses or Sales Related Costs

General exp. (SARE), 3% Sales.	$1,922,897
Royalty, 3% of Production cost	$1,622,815

Total Product Cost

Total Operating Cost	$54,093,847
Annual Cost of Capital	-
Total Product Cost	$54,093,847

Net Income before Taxes

	product lb/hr	sales price	annual sales
Sales	12,565	0.6	$64,096,578
Raw Material			
phenol	13,159	0.15	$16,781,673
ammonia	4,199	0.075	$2,677,492
total raw material cost			$19,459,165

Profit margin (sales - raw materials) $44,637,413

Net Income before taxes (sales - total product cost) $10,002,731

Figure 15 Total Product Cost for Aniline Process

Stranded Costs: This cost is encountered in the utility industry, and it refer to costs that are recovered from the sale of electricity where the price of electricity set by regulations rather than by the market place. Stranded costs are costs where an investment will be less valuable under competition than under regulation. The typical example is an old electrical generating plant where the capital recovery was set presuming that regulations would continue to set the price of electricity for the economic life of the plant. If the market is deregulated or retrofit costs are required to comply with new emissions regulations, a new modern plant would provide power more economically and meet emission requirements. The old plant would be shutdown, and the investment (costs) left in the old plant could not be recovered, i.e., stranded costs. The old plant is said to have become redundant in a competitive environment. Attempts to recover stranded cost include charging all customers in the market area a stranded cost recovery fee and having a government assume a portion of the debt that is then assigned to the public debt. Not all stranded cost occurs in the utility industry. When the refrigerant, Freon, was eliminated from the market by environmental regulations, the plants producing Freon were shutdown, and attempts were made to retrofit them to produce other products.

Sustainable Costs and Credits:

Sustainable development is the concept that development should meet the needs of the present without sacrificing the ability of the future to meet its needs, Brundtland Report, 1987. Sustainable costs are costs to society from damage to the environment by emissions discharged within permitted regulations. Environmental costs are costs to the company that are required to comply with federal and state environmental regulations including permits, monitoring emissions, fines, etc., as described in the AIChE/TCA report (Constable, et al., 2000). The triple bottom line is the difference between (sales and sustainable credits) and (manufacturing costs, environmental costs and sustainable costs). Sustainability credits are the saving from reducing sustainable costs to society by actions from all participants. Details for these costs and credits are given in the following section on Total Cost Assessment.

Economic Evaluation Workflow for Capital, Product and Related Costs

Evaluating total capital cost and total product costs are the needed elements for economic decisions associated with building new plants, retrofitting existing plants, construction scheduling and many other factors required to consider essentially all of the factors that affect the profitability of an investment. In comparing process alternatives, this scope is necessary to ensure that trade-offs between operating cost and capital investments are considered. Computer programming packages are required that provide fast and efficient way for process engineers to estimate plant capital costs and to evaluate plant economics. Estimate of equipment size and cost, utility usage rates, capital investment, operating costs, working capital, start up costs and profitability for any type of process for which there is a continuous flow of material and energy from one processing unit to the next.

In view of the difficulty to maintain and to keep up-to-date cost data and correlations, commercial software for cost estimation is used, and Towler and Sinnott, 2013 provide a description of the computer programs available. They say that the Aspen Technology programs are widely used and are ones that they and we are most familiar. Aspen Economics Evaluation Programs include Aspen Process Economic Analyzer, Aspen Capital Cost Estimator and Aspen In-Plant Cost Estimator. As mentioned previously, the Aspen Capital Cost Estimator program estimates plant capital equipment costs with associated plant bulks using design-based installation models, Aspen Technology, 2013. The Process Economic Analyzer is used to evaluate cash flow and operating costs of a process configuration during design for economic evaluation. Unit operations are mapped into equipment models and costs are calculated using design-based installation models. The Capital Cost Estimator is used to generate cost estimates, budgets and schedules over the project lifecycle with capabilities to optimize control, power, and piping. Investment options including return on investment (ROI) are evaluated, and six regional locations can be used: US, Europe, Middle East, Japan, China, and U.K. The In-Plant Cost Estimator is used for project schedules and detailed budgets with in-plant capital and maintenance projects. Its capability includes evaluating detailed costs for equipment design, CPM construction scheduling and cost analysis.

A detailed description of the workflow to conduct a process economic evaluation was described in Aspen Plus Costing Manual, circa 1988 for the Aspen PLUS costing program, an earlier version of the Aspen programs mentioned above. This workflow description is still relevant and is summarized here.

ASPEN PLUS costing program estimated equipment size and cost, utility usage rates, capital investment, operating costs, working capital, start up costs, etc., and it could be used on a stand-alone basis, or as an integral part of an ASPEN PLUS process simulation. When used on a stand-alone basis, the user entered the information required for the equipment sizing and costing calculations. When used as a part of the process simulation, this information was retrieved from the ASPEN PLUS flow sheet simulation heat and material balance results as indicated in Figure 16.

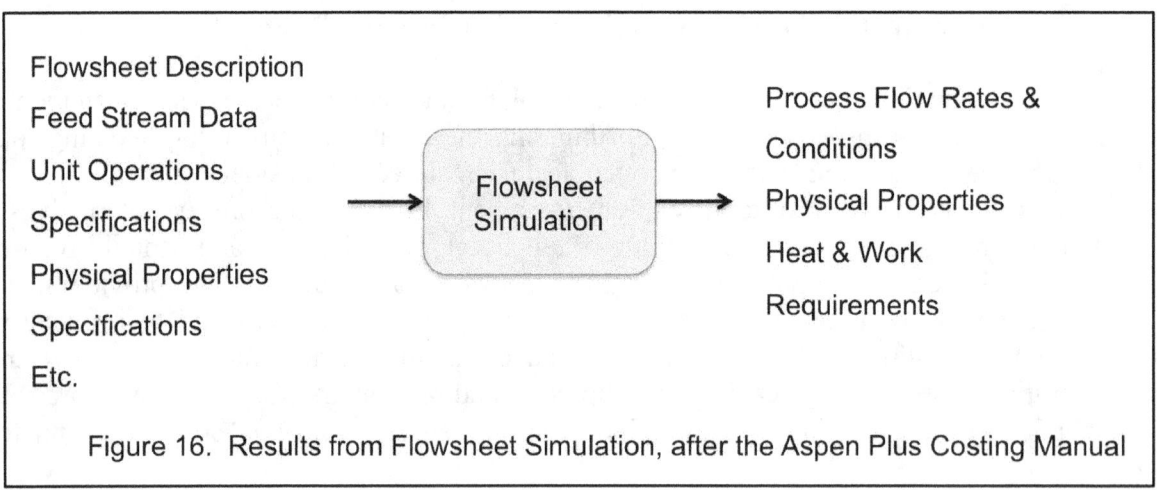

Figure 16. Results from Flowsheet Simulation, after the Aspen Plus Costing Manual

The ASPEN PLUS costing program used an estimating factor method to generate the cost estimates. The estimating factor method has been replaced by the Aspen Icarus technology in the current Aspen Capital Cost Estimator program that is based on and estimates plant capital equipment costs with associated plant bulks using design-based installation models.

As shown in Figure 17, the input data from the flowsheeting program are process flow rates, physical properties, and heat and work requirements were used to size major pieces of equipment, such as heat transfer area, vessel dimensions, temperatures, and pressures, and their costs. The utility use rates for each equipment item are determined and, consumption rates are calculated for utilities such as cooling water, process steam, fuel oil, fuel gas, coal, electricity, and refrigerant.

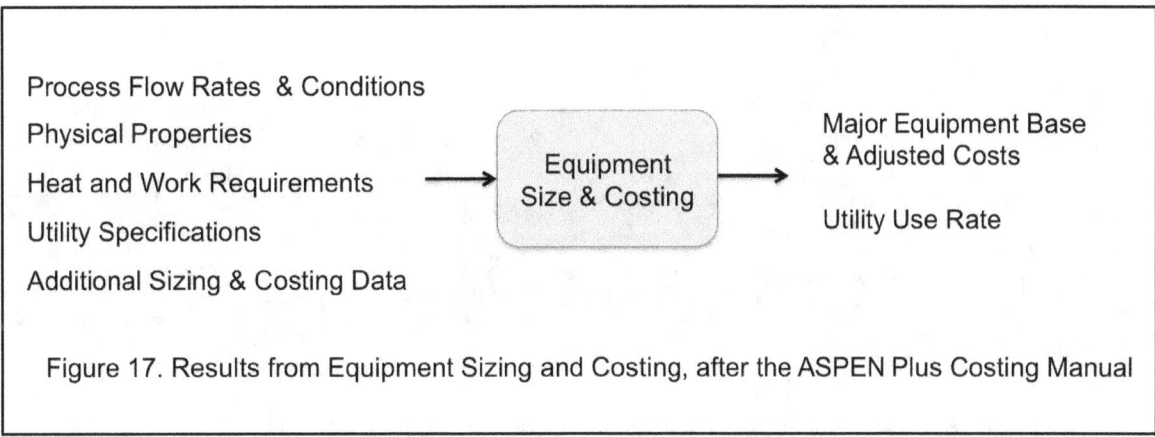

Process Flow Rates & Conditions

Physical Properties

Heat and Work Requirements

Utility Specifications

Additional Sizing & Costing Data

Equipment Size & Costing

Major Equipment Base & Adjusted Costs

Utility Use Rate

Figure 17. Results from Equipment Sizing and Costing, after the ASPEN Plus Costing Manual

Fixed Capital Cost Estimation: As illustrated in Figures 18, in this step the total fixed capital cost is estimated by applying a series of factors to the cost of the major equipment items and by making adjustments for inflation and project location. The installed equipment costs are estimated for each process, utility, and storage unit. Also, other capital costs such as site cost, indirect costs, contingency allowance, etc., are estimated for the plant.

A major factor in the process cost is the installation cost of the process equipment that includes piping, concrete, steel, electrical, instrumentation, insulation, and paint. This installation cost is estimated by choosing the modular method, the unit factor method, or the plant factor method. The modular and unit factor methods estimate the cost of each major equipment item by applying a set of estimating factors compiled for that type of equipment. The plant factor method estimates the cost for the entire plant by applying estimating factors compiled for that type of plant. Of these three methods, the modular and unit factor methods are more accurate than the plant factor method.

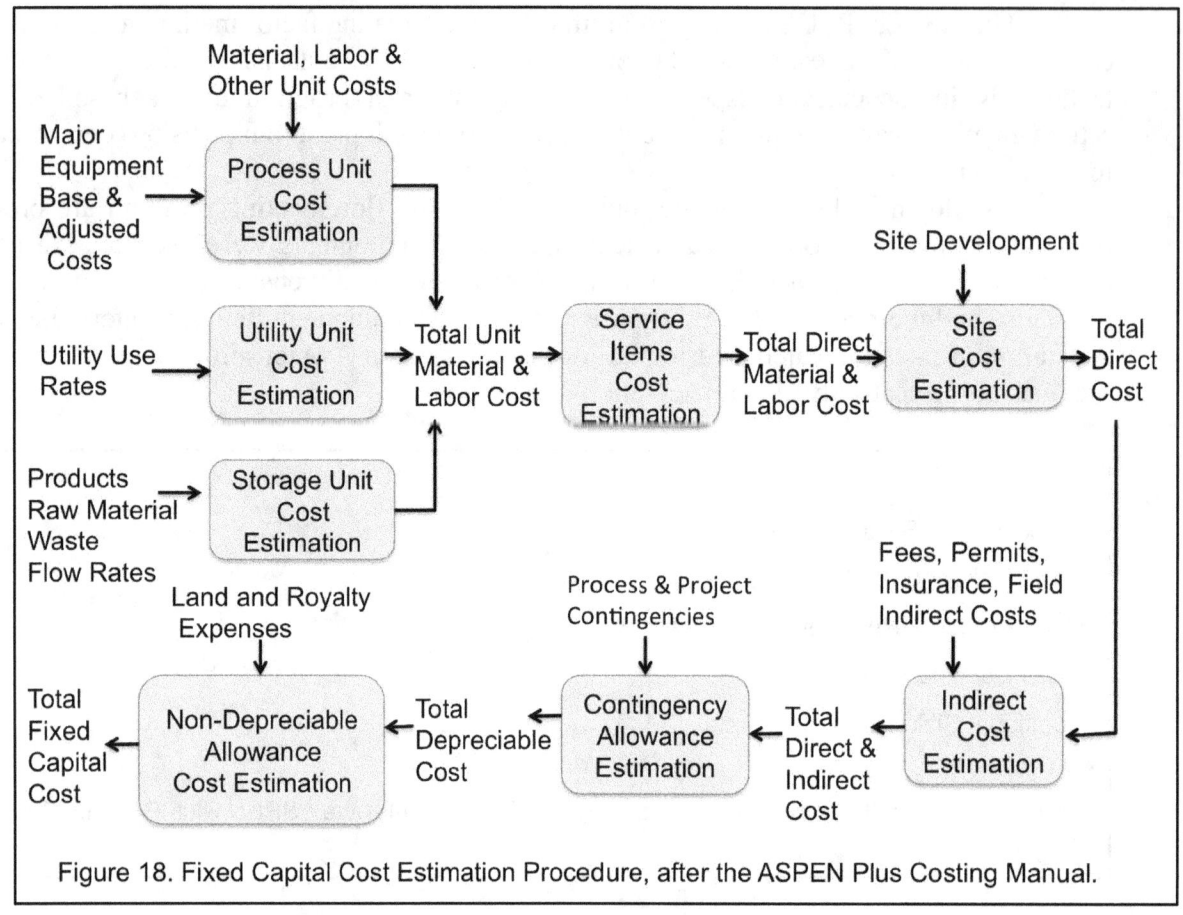

Figure 18. Fixed Capital Cost Estimation Procedure, after the ASPEN Plus Costing Manual.

Utility unit cost estimation is used for utilities generated on-site. The capital cost is estimated, and a user-supplied power law correlation and the utility usage rate can be used.

Storage unit cost estimation is for storage facilities that have not been included as major equipment items or unlisted equipment. The capital cost of a storage unit is estimated, and a user-supplied power law correlation and the volumetric flow rate of the material can be used.

Once the capital cost, and total direct material and construction labor costs for the plant have been estimated, they are adjusted for inflation and project location. This can be done by a cost indexing system.

Operating Cost Estimation: In Figure 19, the components to estimate the operating cost are shown. Operating costs for expenses connected with the manufacturing operation are the total product costs and are divided into manufacturing costs and general expenses as shown in Table 5. These include the cost of raw material, utilities, waste treatment, catalysts, and interest on loans. By product revenue is considered to be a credit against operating cost. The fixed or capital related costs include the cost of operating and maintenance labor and supplies, general and administrative costs, insurance, state and local taxes, and depreciation as shown in Table 5. General expenses or sales related costs are determined are determined from estimating factors as shown in Table 5.

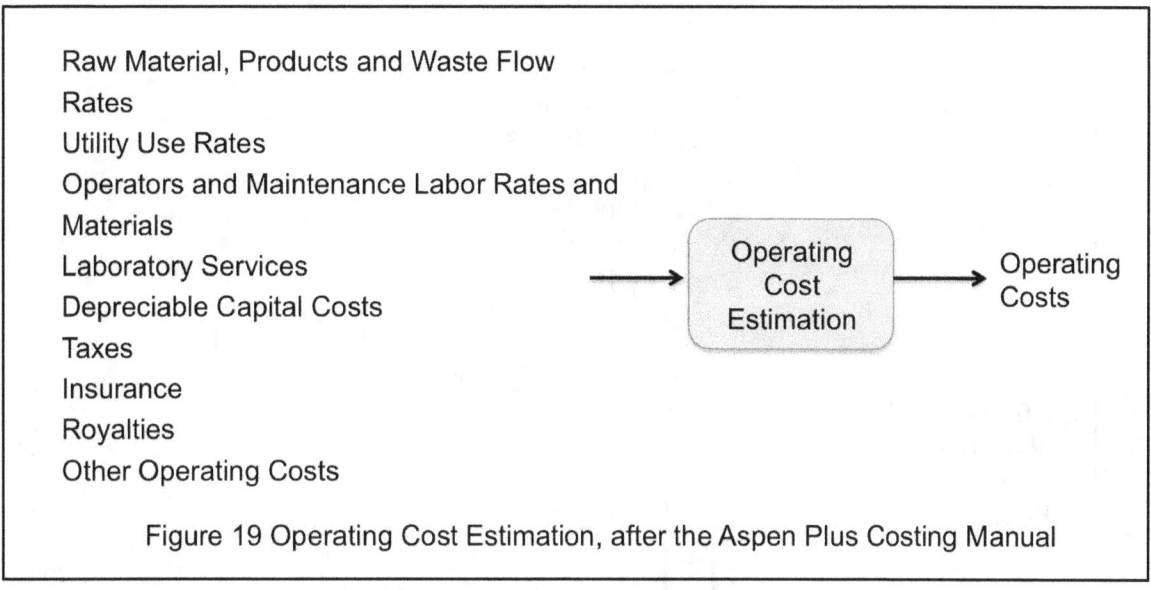

Raw Material, Products and Waste Flow
Rates

Utility Use Rates

Operators and Maintenance Labor Rates and

Materials

Laboratory Services

Depreciable Capital Costs

Taxes

Insurance

Royalties

Other Operating Costs

Figure 19 Operating Cost Estimation, after the Aspen Plus Costing Manual

Typically, two operating cost reports are generated. An annual operating cost summary reports includes annual raw material, utility, waste treatment usage costs, byproduct credits, operating and maintenance material, and labor cost. This report can be generated for 100%, 75%, and 50% of plant capacity. The second report, called the utility usage summary, gives the consumption of each utility.

Economic Evaluation: As shown in Figure 20, costing estimates are generated from a cash flow analysis of the plant. The analysis includes estimates of working capital and start up costs in addition to the fixed capital and operating costs as shown in Table 4. Working capital is estimated from the cost of the various types of inventory, operating cash, accounts receivable, and accounts payable. These items are determined from estimating factors applied to the operating cost. The start-up cost includes the cost of engineering required to achieve commercial production, the cost of inefficiency of start up due to wasted material, and the labor training cost. These items are estimated from factors applied to total plant cost as shown in Table 4. The total investment required to bring the plant into production is calculated as the sum of total capital cost, working capital, start-up costs, etc. as shown in Table 4.

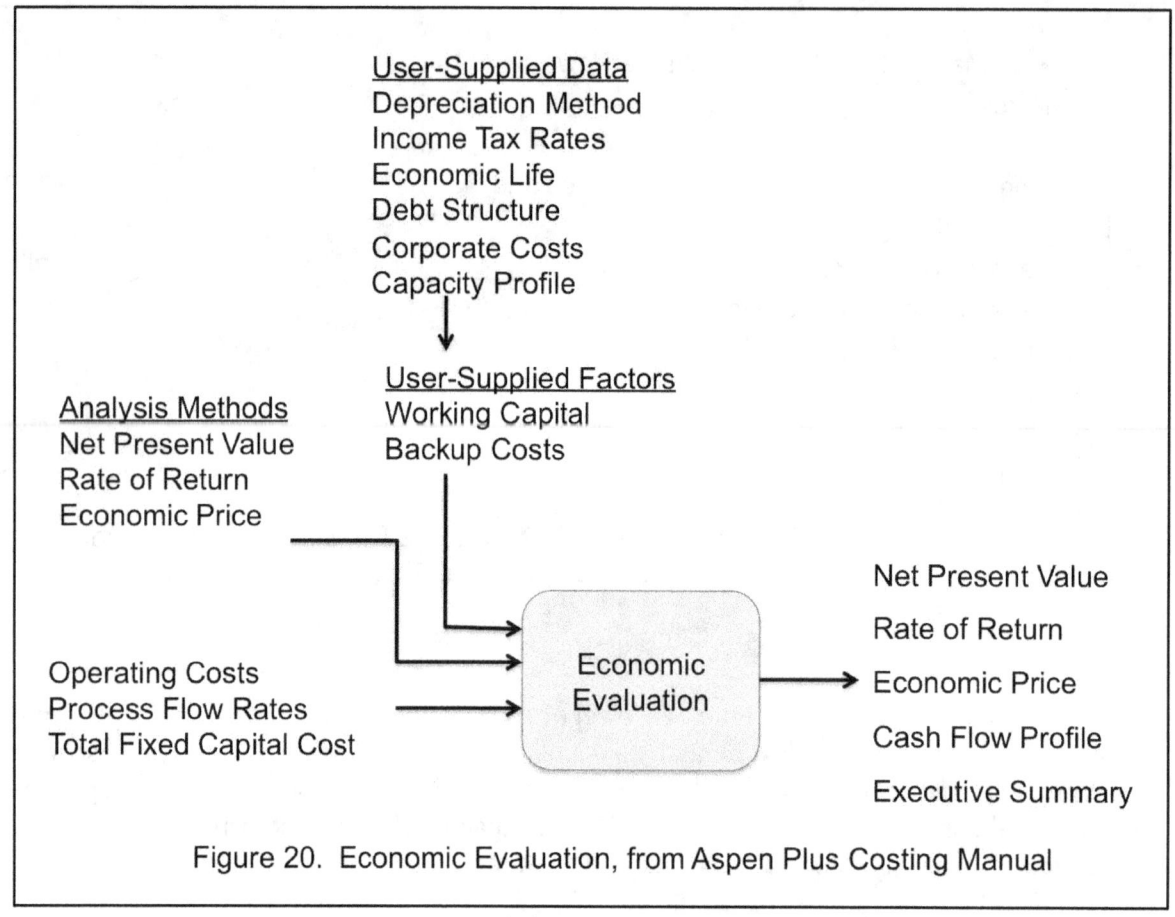

User-Supplied Data
Depreciation Method
Income Tax Rates
Economic Life
Debt Structure
Corporate Costs
Capacity Profile

User-Supplied Factors
Working Capital
Backup Costs

Analysis Methods
Net Present Value
Rate of Return
Economic Price

Operating Costs
Process Flow Rates
Total Fixed Capital Cost

Economic
Evaluation

Net Present Value

Rate of Return

Economic Price

Cash Flow Profile

Executive Summary

Figure 20. Economic Evaluation, from Aspen Plus Costing Manual

Once the total investment is determined, the cash flow profile over both the project and plant operation phases can be calculated. The economic life of the plant and the price profile is specified by the user. The revenue profile is calculated from the production capacity profile specified by the user. The design production quantity can either be retrieved from the flow sheet or specified by the user. Further, the user can also specify various elements that affect the cash flow calculation, such as the structure of debt financing, tax rate, and depreciation. Various IRS-approved depreciation schedules can be incorporated.

The main purpose of calculating the cash flow profile is to evaluate plant profitability using three methods. If the product selling prices are given, an interest rate of return on investment capital can be evaluated that will reduce the investment balance to zero at the end of plant operation. (See Equation 23.) Alternatively, if the interest rate of return is given, the net present value and the economic price can be evaluated. (See Equations 23 and 26.) Details for these methods are described in the next section

The accuracy of the costing system depends on the quality of the data as well as on the judgment of the user. In most cases using the factors method, capital estimates can generate that are better than $\pm$ 30%, while with Aspen Icarus technology accuracies can be of $\pm$ 10%.

Profitability (Economic Decision) Analysis

The bases for profitability analysis used by private corporations are net present value (NPV), rate of return (ROR) and the economic price. Projects are ranked by these three measures of return on investment and compete for the limited capital available for plant improvements and new processes. The **net present value** is the sum of all of the cash flows for the project discounted to the present value, usually using the company's minimum attractive rate of return, MARR; and the capital investment required. The **rate of return** is the interest rate in the net present value calculation that gives a zero net present value. The **economic price** is computed using the sum of the total product cost and the annual capital cost based on the rate of return required by the company that is divided by the product rate. Some background information is needed to evaluate the net present value, the rate of return and the economic price. These include profit, capital expenses, cash flow and the time value of money that are discussed in the next section.

Time Value of Money: Investments mean committing funds in the present with some assurance that a larger amount of money will be returned in the future. This growth in money with is called the **time value of money**. The cost of borrowed money is **interest**, and it is the return for lending money. Also, interest is additional funds that can be obtained when money is put to productive use.

A **simple interest** rate, i, means that only the principal, P, is used in calculation of interest, I, due. Thus:

$$I = iPn$$
(2)

where n is the number of interest periods. It is understood that n and i refer to the same unit of time (year, month, etc.). For example, if $1000 is borrowed for five years at 8% simple interest, the total interest would be:

$$I = (0.08)(\$1,000)(5) = \$400$$

and the total amount, F, due at the end of the loan period is equal to:

$$F = P + I = \$1000 + \$400 = \$1,400$$

Simple interest is rarely used, and usually interest is compounded. **Compound interest** means that interest that has been accrued over the interest period is subject to the interest rate in the next period. For example, if $1000 was borrowed for five years at 8%, compounded annually, the amount of interest due after the first year could be calculated just as simple interest, since there is no compounding in the first year.

$$F_1 = P + iP = P(1+i) \quad F_1 = \$1,000(1+0.08) = \$1080$$

In the second year the interest rate i (= 8%) is then applied to F_1 (= \$1080) if no payment is made at the end of the first year:

$F_2 = iF_1 + F_1$

$F_2 = iP(1+i) + P(1+i)$

$F_2 = P(i+1)^2$

$F_2 = \$1,000(1+0.08)^2 = \$1,166$

In the third year, the procedure is repeated:

$F_3 = iF_2 + F_2$

$F_3 = iP(i+1)^2 + P(i+1)^2 = P(i+1)^2(i+1)$

$F_3 = P(i+1)^3$

$F_3 = \$1000(1+0.08)^3 = \$1,260$

The amount to be repaid in years four and five would be:

$F_4 = P(i+1)^4 = \$1,361$

$F_5 = P(i+1)^5 = \$1,469$

In general, the formula to compute the compound interest amount, F_n, to be paid at the end of n time periods with an interest rate of i is given by the following equation.

$F_n = P(1+i)^n$
(3)

This equation gives the *future worth*, F_n, of an amount P in the present (*present value*).

Compounding can be calculated for more than once a year, e.g. quarterly, monthly, daily, even continuously. However, interest rates are usually quoted as an *annual nominal interest rate,* r. If m is the number of times the interest is compounded between payment, then the *annual effective interest rate*, i_e, is given by the following equation (Sepulveda, et al. 1984)

$i_e = (1+r/m)^m - 1$
(4)

For example, using Equation 4 for a nominal interest rate r = 0.08 compounded quarterly (m = 4) the effective interest rate is $i_e = 0.0824$ from Equation 4. The value of i_e is used in Equation 3 to evaluate F_n given P; and for P equal to $1,000, the value of F_5 is $1,486 for five years.

As shown by Equation 3, money, having an ability to earn interest, has its value increase over time, thus the term *time value of money*. The *future worth*, F_n of an amount of money P is given by Equation 3. Also, P is called the *present value* of an amount of money whose future worth is F_n that is available in n time periods at an interest rate of i. Note that an amount of money F_n available in n time periods in the future is worth less than the same amount of money available in the present. The amount F_n decreases in value is shown by rearranging Equation 3 to have:

$$P = F/(1 + i)^n$$
(5)

In the example, $1,469 received in five years is only worth $1,000 today if the interest rate is 8%.

Inflation: Inflation is defined as the overall general upward price movement of goods and services in an economy according to the Bureau of Labor Statistics (BLS) of the U. S. Department of Labor, and BLS has various indexes that measure different aspects of inflation (www. Bis.gov/bls). The commonly quoted inflation rate is the change in the Consumer Price Index (CPI-U) from a year earlier (InflationData.com). The inflation rate is shown in Figure 21 from 1990 to 2012 using BLS data which goes back as far as 1913 (www.bls.gov/cpi).

If inflation is at a fractional annual rate, f, both the present value, P, and the interest, I, are reduced by the inflation rate and that reduces the future worth, F, i.e.,

$$F = (1 + i)P - f F \qquad \text{or} \qquad F = [(1 + i)/(1 + f)]P$$

or another way gives the same result.

$$F = P/(1 + f) + I/(1 + f) = P/(1 + f) + iP/(1 + f) = [(1 + i)/(1 + f)]P$$
(6)

For a constant interest rate and constant inflation rate, this equation can be written as the following for n years:

$$F = P [(1 + i)/(1 + f)]^n$$
(7)

For example, if the inflation rate was 6% (f = 0.06) for the five year period (n = 5) when P = $1,000 was invested at an interest rate of 8%, the future worth is only $1,098 compared to $1,469 not considering inflation.

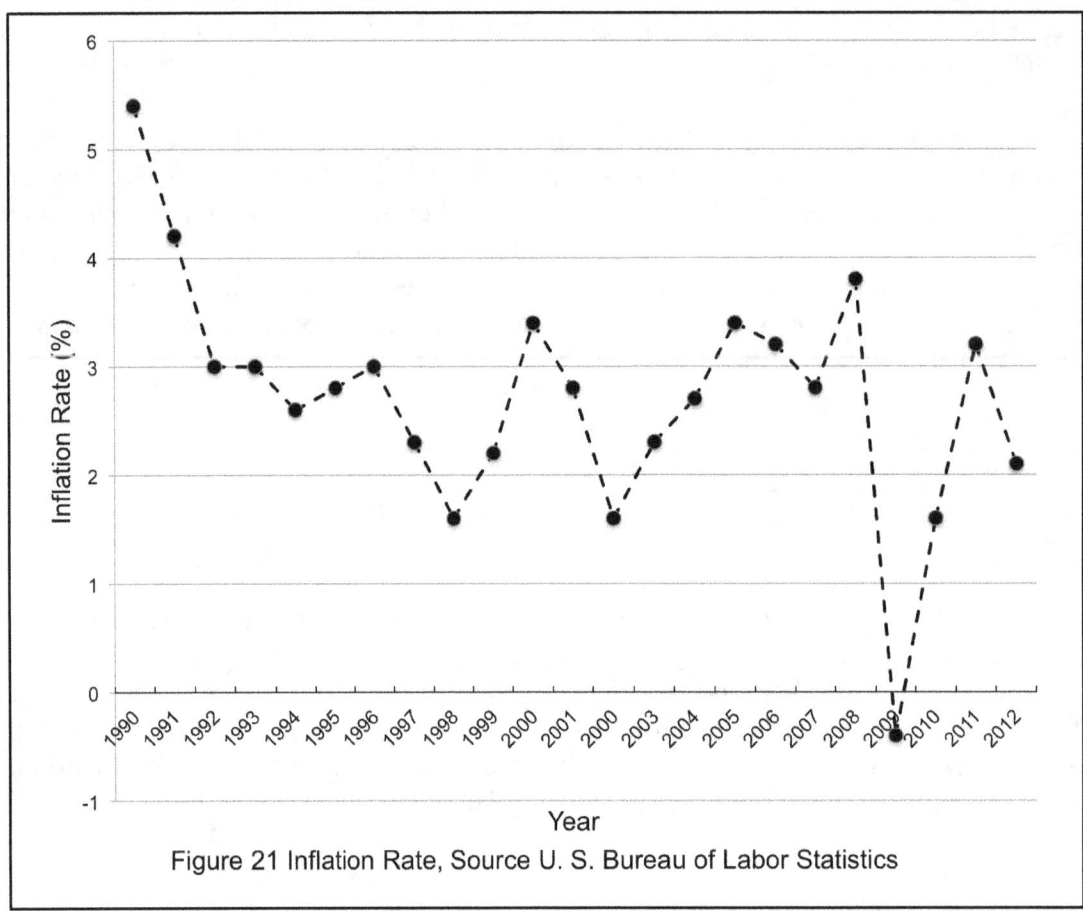

Figure 21 Inflation Rate, Source U. S. Bureau of Labor Statistics

For a constant inflation rate, f, expressed a fraction, over a period of n years, the future cost of a commodity, F_c, increases in relation to the present cost, P_c, by the following equation (Sepulveda, et. al., 1984).

$$F_c = P_c(i+f)^n \qquad (8)$$

Taxes: A detailed discussion of taxes by Federal, state and local governments and the related topic depreciation is given in a subsequent section. As an introduction, consider interest as taxable income, and interest earned in a given year is subject to taxes for that year. If the interest period is the same as the tax period, then the interest earned during this period is iP, and the tax due on this earned interest is tiP where t is the tax rate. The funds retained after taxes are $(1 - t)iP$. For companies t can be as much as 0.38, and the net return after taxes is interest minus taxes, $I - T = iP - tiP = (1-t)iP$. Consequently, the future worth, F, after one year after having paid taxes is:

$$F = P + (1-t)iP = [1+(1-t)i]P$$
$$(9)$$

Taxes and Inflation: If inflation proceeds at rate f during this year, the above equation can be modified to include the tax and inflation rates as:

$$F = P/(1 + f) + (1 - t)iP/(1 + f) = \{[1 + (1 - t)i]/(1 + f)\}P$$

With some rearrangement, the above equation can be put in the following form.

$$F = \left[1 + \frac{(1-t)i - f}{1+f}\right]P \qquad (10)$$

Then Equation 10 has the form of Equation 3 by defining a composite interest rate, i_c, that includes inflation and taxes as:

$$i_c = \frac{(1-t)i - f}{(1+f)} \qquad (11)$$

Now, the effects of interest, inflation and taxes can be included in the evaluation of the future worth F knowing the present value P by the following equation (Sepulveda, et. al., 1984).

$$F = P(1+i_c)^n \qquad (12)$$

For a tax rate of 0.20 with an inflation rate of 0.06, the future worth of $1,000 invested at an interest rate of 8% for five years is now only $1,019 compared to $1,469 not considering inflation and taxes.

The intervals for the interest, inflation and tax rates in Equation 12 must all be the same, and the rates must be constant. If this is not the case, the same procedure is used, but each interval has to be evaluated separately and sequentially.

Risk: The effect of risk can be approximated in this analysis by replacing the interest rate i in Equation 11 with the sum $(i+i_r)$. Here i_r represents an addition to the cost of financing a project when there is more risk involved than is normally expected. Additional information is given in a subsequent section on Risk Analysis

Cash Flow: To this point we have been discussing the results of investing a single sum of money. However, money received from the sale of products and spent on manufacturing costs by a company occurs on nearly a continuous basis; and income is reinvested daily by the company. To be able to analyze the income and expenditures for a company, it is convenient to select a time interval, typically a year, to perform the evaluations. The *cash flow* is determined for the period that is the difference between all of the funds received and all of the funds disbursed. It is convenient to represent this net annual cash flow on a diagram, and this is especially useful for project evaluations. A *cash flow diagram* will show what is expected to take place over the life of a project where the horizontal axis represents time intervals, and the vertical axis represents the amount of cash flow in each of these times.

A simple cash flow diagram is shown in Figure 22 where the negative cash flows in the first two years represent a net loss for the project. After that, all flows are positive, meaning a net gain in each of those three years. This diagram could apply to a company that planned to purchase a new generator that would reduce utility bills by $4,000 per year for six years, and the generator would be purchased with two payments of $9,000 in the first two years. Then at the end of the sixth year the generator could be sold at a salvage value of $1000 that is added to the income received in the sixth year. Receipts are represented by arrows directed up, and disbursements directed down where the length of the arrow is proportional to the magnitude of the cash flow. Expenses incurred before the start of the project are called sunken costs.

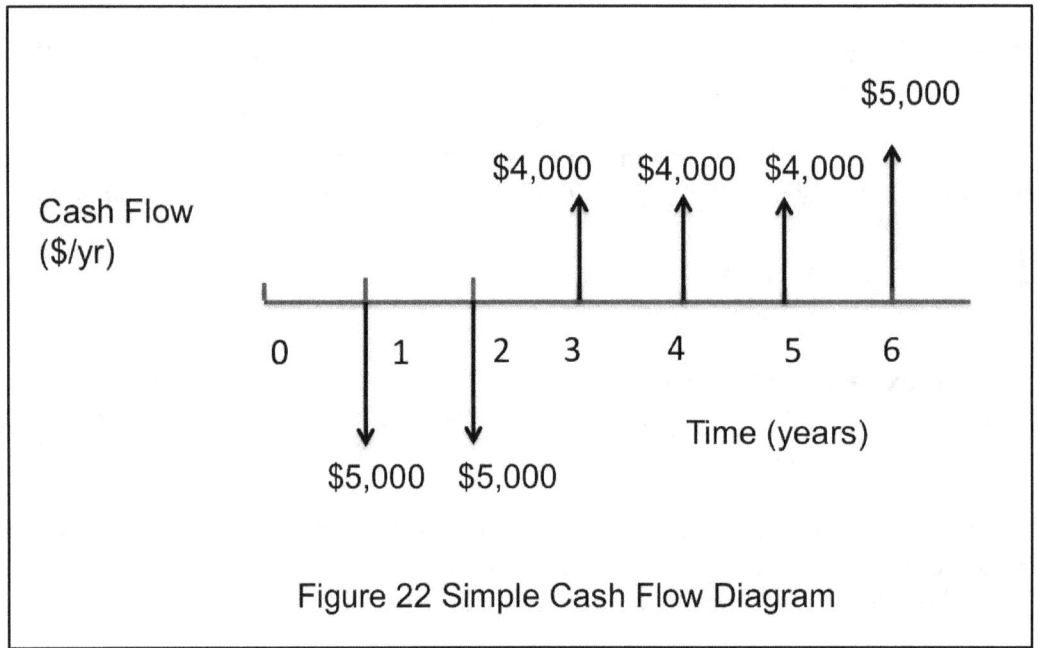

Figure 22 Simple Cash Flow Diagram

This example illustrates the **end of year convention** that assumes that income and disbursements are made at the end of each year. Although receipts and payments are usually made throughout the year, it greatly simplifies calculations; and, practically, it poses no problem, since it will not lead to errors in choices between alternatives. Exceptions to the year-end convention are initial project costs (purchase costs), trade-in allowances, and other cash flows associated with the project.

A more realistic cash flow diagram is shown in Figure 23 for a typical projection of the net annual income for the estimated economic life of a new plant. To construct this diagram many items must be estimated. These include demand for product, plant capacity to meet the demand for the company's planned penetration into the market, selling price, cost of raw materials, direct fixed capital for design and construction of the plant, allocated and working capital, land costs and out-of-pocket expenses. As shown in the diagram, there is a net loss through the sixth year, and there is a net profit from the seventh year through the end of the plant's planned life in the fifteenth year. The **break-even point** is shown between the sixth and seventh year when the revenues match expenses, and the profit is zero. This gives the **break-even capacity** that is the production rate when all of the costs, excluding depreciation, are equal to the sales realized.

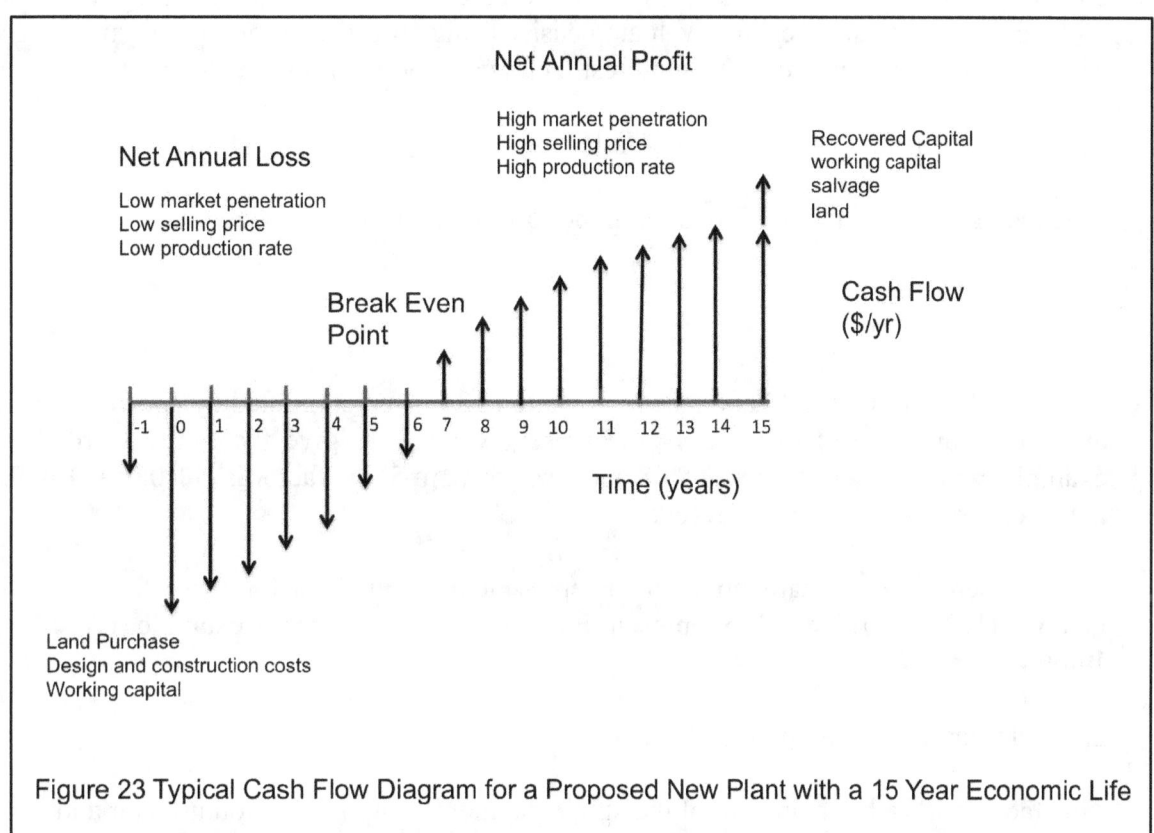

Figure 23 Typical Cash Flow Diagram for a Proposed New Plant with a 15 Year Economic Life

The economic life is estimated based on the length of time the plant can be operated profitably. New, more efficient technology to produce the product, new environmental restrictions and a new product from another process that displaces the current product will end the economic life of a plant. An example of new technology is the replacement of the lead chamber process with the contact process for the manufacture of sulfuric acid. New environmental restrictions on the use of fluorocarbons for refrigerants have caused these plants to be shut down. In the field of polymers, there has been a constant replacement of new plastics replacing existing ones starting with the first plastic, Bakelite.

The break-even point and the payback period are two ideas that are used in **break-even analysis** to compare different projects. Others include determining the useful life for alternate pieces of equipment and the capacity utilization of alternative pieces of equipment in terms of time used per year. Detailed discussions of these topics are given by De Garmo, et al., 1988, Garrett, 1989 and Turton, et.al., 1998.

Interest Factors: There are several interest factors that are routinely used in economic decision analysis to evaluate the present value P, future worth F, uniform payment (receipt) A, and uniform gradient G. These equations require a constant interest rate i over a series of uniform time intervals n. Values for these factors can be found in tables of compound interest factors. Also, some electronic calculators incorporate these calculations with special keys, and they are evaluated easily using spreadsheet programs, such as Excel. The cash flow diagrams for theses interest factors are shown in Figure 24.

1. Single-Payment, Compound-Amount Factor:

This factor, F/P, is Equation 3 written in the following form.

$$\frac{F}{P} = (1 + i)^n \tag{13}$$

Values of the right hand side of Equation 13 can be computed by specifying i and n, and these can be used to multiply the present value P to give the future worth F. For example, with an interest rate i = 0.08 and five years, n=5, the factor is $(1+.08)^5 = 1.469$, and the future worth of $1,000 is $1,469.

There is a standard notation to represent the equations for these factors, and the notation (F/P, i%, n) is used to represent Equation 13. Thus, for the example (F/P, 8%, 5) = 1.469.

2. Single-Payment, Present-Worth Factor:

This factor P/F is the reciprocal of the single payment compound-amount factor and is given by the following equation.

$$\frac{P}{F} = (1 + i)^{-n} \tag{14}$$

The quantity $(1 + i)^{-n}$ is called the **discount factor**, also. The future worth F can be multiplied by this factor to determine the present value P. For example, with interest rate i=0.08 and n=5, the value of the factor is 0.68058, and the present value of F = $1,469 is $1,000. Here (P/F,8%,5) = 0.68058.

3. Uniform-Series, Compound-Amount Factor:

This factor F/A gives the future worth, F, of a uniform series of equal payments or receipts, A, that are made over n years earning an interest rate i. The equation for this factor is given below. (See Figure 24.)

$$\frac{F}{A} = \frac{(1+i)^n - 1}{i} \tag{15}$$

This equation was obtained by showing that the total amount accumulated is the sum of a geometric progression, i.e., $F = A(1+i)^{n-1} + A(1+i)^{n-2} + ... + A(1+i) + A = A[(1+i)^{n-1} + (1+i)^{n-2} + ... + (1+i) + 1] = A[(1 + i)n - 1]/i$.

A uniform annual series of payments can be multiplied by this factor to determine their future worth F. For example, if \$1,000 is invested annually for five years, n=5, at an interest rate of i=0.08, the value of this factor is 5.867; and the future worth of these funds is \$5,867. Here (F/A,8%,5) = 5.867.

4. Uniform-Series, Sinking-Fund Factor:

This factor A/F is the reciprocal of the uniform-series, compound amount factor and is given by the following equation (Sepulveda, et. al., 1984).

$$\frac{A}{F} = \frac{i}{(1+i)^n - 1} \tag{16}$$

This factor provides a means to compute the value of a uniform series, A, to have a total amount F accumulated after n years. For five years (n=5) and an interest rate of i=0.08, the factor is 0.17045; and using F=\$5,8767 from above, then A=\$1,000. Here (A/F,8%,5) = 0.17045.

5. Uniform-Series, Capital-Recovery Factor:

This factor A/P gives the uniform series value A that depletes an amount of money P over n years with an interest rate of i. This equation can be obtained by multiplying Equations 13 and 16, i.e. A/P = (F/P)(A/F), and the result is (Sepulveda, et. al., 1984).

$$\frac{A}{P} = \frac{i}{1-(1+i)^{-n}} \tag{17}$$

For five years (n=5) and an interest rate of i=0.08, this factor is 0.25046, and for an amount P = \$3,993, the value of A is \$1,000, i.e. five payments of \$1,000 could be distributed. Here (A/P, 8%, 5) = 0.25046.

6. Uniform-Series, Present Worth Factor:

This factor P/A is the reciprocal of the uniform-series, capital-recovery factor, is used to compute the principal needed to assure a uniform series of payments for n years at interest rate i. The equation is:

$$\frac{P}{A} = \frac{1-(1+i)^{-n}}{i} \tag{18}$$

This equation has a value of 3.993 for n=5 and i=0.08; and for A=\$1,000, the value of P is \$3,993. Here (P/A, 8%,5) = 3.993.

7. Gradient Series Factor:

This factor A/G is for an initial series value of A_o and each succeeding year A_o is increased, first by an amount G then 2G as shown in Figure 24. At year n the series value is $A_o + (n-1)G$. The amount accumulated after n years is F, obtained by summing a geometric progression, (Sepulveda, et. al., 1984).

$$F = A_0 \left[\frac{(1+i)^n - 1}{i} \right] + \frac{G}{i} \left[\frac{(1+i)^n - 1}{i} \right] - \frac{nG}{i} \tag{19}$$

The amount, F, can be converted into a series of uniform payments, A, by the following equation.

$$\frac{A}{G} = \frac{1}{i} - \frac{n}{(1+i)^n - 1} \tag{20}$$

This equation is used with the gradient-series, present worth factor, P/G, to determine the present worth of the series where P/G = (A/G)(P/A), and (P/A) is given by Equation 18. The use of these factors will be illustrated in a subsequent example.

The cash flow diagrams for these factors are shown in Figure 24. This figure summarizes the interest factors, also.

Minimum Attractive Rate of Return (MARR): Private corporations require a minimum attractive rate of return (MARR) before considering investing in a project. This is an interest rate that may be a project-specific number, but it usually reflects the average return on investment for a particular corporation. Determining the appropriate MARR is a corporate policy matter. However, in an economist's point of view, an investment is attractive as long as the marginal rate of return is equal to or greater than the marginal cost of borrowed capital. Corporations and their investors usually require a substantially higher return than that which could be obtained by simply investing in a money market account.

The source of funds is a consideration when choosing a value for MARR. A private corporation can use funds from the owners, usually through the sale of stock, or from profits that are fed back into the corporation, or from capital recovery. The *cost of equity capital* is related to a company's policy on debt financing. High leverage makes equity capital (money from stock sales) in a business more risky, since equity capital must sustain any losses first before debt capital. Hence, investors require a higher return to compensate for the risk. Also, the minimum attractive rate of return (MARR) generally represents the opportunity cost of money, since funds expended in one project are unavailable for others.

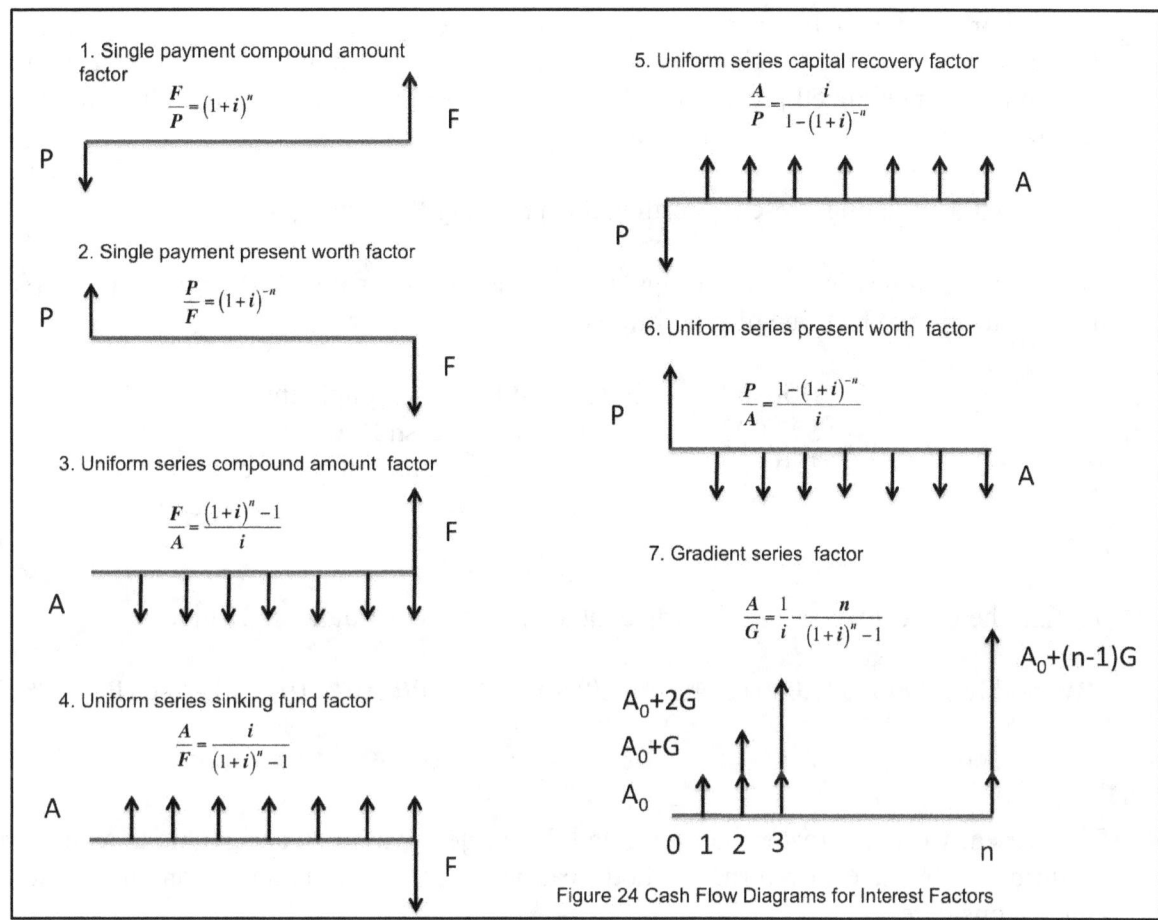

Figure 24 Cash Flow Diagrams for Interest Factors

Methods for Project Analysis

The present worth and annual worth methods of project analysis all assume estimations have been made regarding initial investment, cash flows, acceptable MARR, and the possibility of salvage value on equipment after the project has been completed. The equivalent uniform annual series equation used with these methods is described, also.

Present Worth Method: This method of economic decision analysis converts cash flows, CF_j, into an equivalent present worth or value for a certain minimum attractive rate of return as shown by Equation 21 to evaluate the present worth (PW).

$$PW = \sum_{j=0}^{n} CF_j(P/F) = CF_0 + CF_1(P/F, i\%, 1) + ...+ CF_n(P/F, i\%, n) \qquad (21)$$

where (P/F) is evaluated by Equation 14 for the interest rate i and year j for the cash flow CF_j.

61

For costs, as when a necessary expense is being planned (e.g., the installation of a new cooling system), then the present worth of costs is minimized. For profit, MARR is used, and the maximized present worth has to be greater than zero. The following example illustrates evaluations using present worth.

Example 5. Evaluating Present Worth and Comparing Projects

Consider the following series of payments and profits from ABC Pipes Inc.'s plan to institute a new product line of pipefittings. The cash flows are:

$150,000   initial cost for new equipment
$22,000   yearly after tax cash flow
$40,000   maintenance of equipment in year 10
$30,000   salvage value of equipment in 20 years

To find the present worth of this project at a 10% MARR, Equation 21 gives:

PW = -$150,000 + $22,000 (P/A,10%,20) - $40,000 (P/F,10%,10)+ $30,000 (P/F,10%,20)

    = $9,238

The present worth is greater than zero, and this project would be acceptable to a corporation requiring a 10% rate of return. A modification of this example shows another option that can be considered.

Instead of manufacturing the pipefittings, ABC Pipes Inc. is considering expanding its existing product line of pipes with a new state-of-the-art plastic. The cash flows for this project are:

$72,000   initial investment
6,000   profit in year 1
6,500   profit in year 2, and the profit continues to increase $500 every year for 20 years

Assuming zero maintenance and no salvage value, the present worth for this project is:

PW   = -72,000 + 6000 (P/A,10%,20) + 500 (P/G,10%,20)
     = -72,000 + 6000(8.514) + 500(55.407)
     = -72,000 + 51,084 + 27,704
     = $6,787

Both courses of action would be acceptable to ABC, but if funds were not available for both, other considerations need to be addressed. The pipefittings project is more profitable; and, all things being equal would be the clear choice. Yet all things are not equal. If the lower initial investment for the plastics would leave funds for another, even more profitable project, then that too must be evaluated.

Another concern is competition. What if the project lives of the two alternatives were unequal? The two would not be comparable if the equipment for plastics would have to be retired in ten years. In instances when the analyst needs to compare projects with unequal lives, engineering judgment is required. In this instance, both could be considered over 10 years, or over 20 years. If additional new equipment for $72,000 would be purchased in year 10, the new plastics alternative would then look like this:

$$PW = 72,000 + 6000 \ (P/A,10\%,20) + 500 \ (P/G,10\%,20) - 72,000 \ (P/F,10\%,10)$$
$$= -\$20,969$$

An accurate estimation for project life is essential in economic analysis like this one.

Annual-Worth Method: This method, also called equivalent uniform annual disbursements, converts uneven cash flows into their equivalent uniform annual values using the equivalent uniform annual series as described below. The results are the same as in the present worth method; that is, the indicated best choice between alternatives will always be the same, but annual worth may be easier to grasp.

Equivalent Uniform Annual Series: This is an equation that converts the present worth, PW, from a series of cash flows to a series of uniform annual payments. The present worth, PW, for the series of cash flows is computed using Equation 21, and (P/F) is evaluated by Equation 14 for the interest rate i and year j for the cash flow CF_j. Then the present worth is converted to a uniform series (EUAS) using Equation 17 for (A/P), the uniform-series, capital-recovery factor, i.e.:

$$EUAS = PW(A/P) = PW\{i/[1 - (1 + i)^{-n}]\} \tag{22}$$

When the cash flows are negative representing costs, then the above equation is called the *equivalent uniform annual cost* (**EUAC**), *annual cost* (**AC**) or *capital recovery* (**CR**). The following example illustrates this procedure.

Net annual savings can be used for project evaluation. The cost of equipment for a proposed project is converted to an equivalent uniform annual cost (EUAC) using the interest rate for the minimum attractive rate of return and the project life. This gives the annual cost of capital for the equipment. If this annual cost is less than the projected saving from the use of the equipment, then the project is economically attractive.

Example 6. Applying EUAC to Compare Potential Projects

A company needs additional warehouse space for product as a result of plant expansion. The options include constructing a prefabricated steel building, a tilt-up concrete building or renting space. The steel building has a cost of $150,000 and a service life of 25 years with annual maintenance and property taxes of $6,000 per year. The concrete building has a cost of $200,000 and a service life of 50 years with annual maintenance and property taxes of $4,000. Both buildings have no realizable salvage value, and the company uses a 15% minimum attractive rate of return. The company can rent suitable space for $32,000 per year. Basing the decision for additional warehouse space on the equivalent uniform annual cost, should the company: (A) Construct the steel building, (B) Construct the concrete building or (C) Rent warehouse space.

Using Equation 22, the equivalent uniform annual cost (EUAC) for the buildings is given by:

$$EUAC - -P(A/P,i\%,n) - [\text{maintenance and taxes}]$$

where (A/P,i%,n) is the uniform-series, capital-recovery factor, and P is the cost of the building.

Steel Building: EUAC = -150,000(A/P,15%,25) - 6,000
EUAC = -150,000(0.15470) -6,000 = -$29,205

Concrete Building: EUAC = -200,000(A/P,15%,50) - 4,000
EUAC = -200,000(0.15014) - 4,000 = -$34,028

Comparing the above annual rates with renting at $32,000 per year, the best decision is to build the steel building for $29,205 equivalent uniform annual cost.

The following example illustrates another way to use the equivalent uniform annual cost to determine the best alternative for an investment decision.

Example 7. Applying EUAC to Evaluate a Potential Capital Investment

Refractory bricks lining a furnace have an installed cost of $35,000 and last six years. The furnace must be partially relined at the end of three years at a cost of $12,000. A new high temperature refractory material has been developed, and test results show that a lining with this material will last 15 years with no intermediate repair costs. What is the largest annual cost that can be justified economically for using this new material? The company uses a 25% minimum attractive rate of return.

The equivalent uniform annual cost (EUAC) for the current and new refractory materials will be equated to determine the maximum initial cost of the new material using the company's minimum attractive rate of return of 25%.

Current material: initial cost = $35,000 lasting 6 years
Repair cost = $12,000 after 3 years
EUAC   = 35,000(A/P,25%,6) + 12,000(P/F,25%,3)(A/P,25%,6)
         = 35,000(0.33882) + 12,000(0.51001)(0.33882) = $13,941 per year

New material: The maximum equivalent uniform annual cost for new material lasting 15 years = $13,941. Let P = initial cost of new material and

13,941 = P(A/P,25%,15)
13,941 = P(0.25912)
P = $53,801

Thus, $53,801 is the largest annual cost that can be justified for this material.

Discounted Cash Flow Calculations: The net present value, sometimes called discounted cash flow (Holland et. al., 1983) (DCF), is an estimate of profitability of a project. It has the advantage that the net present value for several projects can be added to obtain the net present value for all of the projects.

The rate of return is sometimes called the discounted cash flow rate of return (DCFRR) and the internal rate of return (IRR). The rate of return has the advantage of being used to compare projects with alternate uses of money that have rates of return, such as bonds and certificates of deposit.

There are numerous similar measures of profitability, but all of these are variations of net present value and rate of return, except the payback period or payout time that has the flaw of neglecting the time value of money. One example is the net rate of return (NRR) that is the net present value divided by the product of the capital investment at year zero and the product life expressed as a percentage (Ward, 1986).

Net Present Value: To evaluate the net present value, NPV, the net annual cash flows, CF_j, are evaluated using the following mnemonic equation using the nomenclature from Table 1.

Net annual cash flow =	Sales -	Total product -	Taxes -	Annual capital
after taxes, CF_{xt}	S	cost, C_T	T	expenditures, C_{cap}

From Table 1: $CF_{xt} = S - C_T - T - C_{cap}$

These cash flows are used in the following equation to compute the net present value (NPV) where CF_0 is the initial capital investment for the project (a negative cash flow).

$$NPV = -CF_0 + \sum_{j=1}^{n} CF_j(1+i)^{-j} \qquad (23)$$

The net present value, NPV, is the present worth, PW, with CF_0 specified as the total plant cost.

To determine the net present value, the interest rate, i, usually the minimum attractive rate of return, and the number of years, n, for the project are specified. There is no assumption about the signs of the cash flows, CF_j, but the equation has the initial cash flow, CF_o, being negative to represent the initial capital investment.

Referring to Figure 23 and Equation. 23, the individual annual cash flows would be discounted to the present and combined with the capital investment to estimate the net present value for the proposed new plant. The following simple example illustrates the calculation of the net present value.

Example 8. Evaluating Net Present Value

A straight-run fuel oil stream in a refinery can be converted using hydrocracking to a high-octane fuel for blending into premium gasoline. A proposal has been made to add a 15,000 bbl/day unit at a capital cost of $71.0 million. The annual net profit in million dollars is given below for the estimated life of the hydrocracking unit. The net present value is to be evaluated for interest rates of 15% and 25%, and the profitability compared. These results are shown in the following table.

End of Year n	Annual Net Profit, F	P/F(15%) $(1.15)^{-n}$	Present Value	P/F(25%) $(1.25)^{-n}$	Present Value
1	32.0	0.8695	27.83	0.8000	25.60
2	28.0	0.7561	21.17	0.6400	17.92
3	22.0	0.6575	14.47	0.5120	11.26
4	17.0	0.5718	9.72	0.4091	6.96
5	15.0	0.4972	7.46	0.3277	4.92
Total	114.0		80.64		66.66

Computing the net present value gives:

$$NPV(15\%) = -71.0 + 80.84 = 9.64 \quad NPV(25\%) = -71.0 + 66.66 = -4.34$$

The investment is marginally attractive with a positive net present value if funds are available at 15%, but the project is not considered with a negative net present value for funds available at 25%.

A convenient form of Equation 23 is obtained if all of the cash flows, CF_j, are equal. By using Equation 18 gives.

$$NPV = - CF_o + A\,[1 - (1+i)^{-n}\,]/i = - CF_o + A(P/A,\ i\%,n) \tag{24}$$

where A is the uniform cash flow in the equation.

Example 9. Evaluating Net Present Value with the Gradient Series Factor

Two projects are competing for an oil company's capital improvement funds. One is an additional distillation column for improved product quality, and the other is a new lubricant packaging system to reduce product-packaging costs. The capital investment is $120,000 for each one. The cash flow for the distillation column is $38,000 for the first year, and then it declines by $4,000 for each subsequent year for the 10-year life of the project. The cash flow for the packaging system is $5,000 for the first year, and then it increases by $4,000 for each subsequent year for the 10-year life of the project. For a 15% minimum attractive rate of return compute the net present value for each proposed project.

For both of the oil company's projects:

Capital investment, CF_o = $120,000; i = 15% and n = 10 years.

Distillation column cash flows: year 1 CF_1 = $38,000, years 2 - 10 - declines by $4,000.

Packaging system cash flows: year 1 CF_1 = $5,000, years 2 - 10 - increases by $4,000.

The following equation gives the net present value (NPV) for these projects.

$$NPV = CF_o + CF_1(P/A,i\%,10) + G(A/G,i\%,10)(P/A,i\%,10)$$

where G is the gradient of the cash flow, (A/G,i%,n) is the gradient series factor given by Equation 20 and (P/A,i%,n) is the uniform series, present-worth factor given by Equation 18.

For the distillation column:

$$
\begin{aligned}
NPV &= -120,000 + 38,000(P/A,15\%,10) - 4,000(A/G,15\%,10)(P/A,15\%,10) \\
&= -110,000 + 38,000(5.0188) - 4,000(3.3832)(5.0188) \\
&= \$2,796
\end{aligned}
$$

For the packaging system:

$$
\begin{aligned}
NPV &= -120,000 + 5,000(5.0188) + 4,000(3.3832)(5.0188) \\
&= -\$26,988
\end{aligned}
$$

The distillation column is a potentially attractive investment with a positive net present value, but the lubrication packaging system is not with a negative net present value.

Example 10 Comparison of Net Present Value and Net Annual Savings

A pyrolysis furnace is used to thermally crack ethylene dichloride to vinyl chloride monomer. A diagram of the furnace shown in Figure 25, and it is used to heat 100 million pounds per year of ethylene dichloride from 261°F to 930°F, and then the material is cooled in a heat exchanger, as described by Mr. D. O. Hutchinson of Ethyl Corporation. It is proposed to apply energy integration as shown in the diagram by using the material exiting the furnace to heat the feed to the furnace.

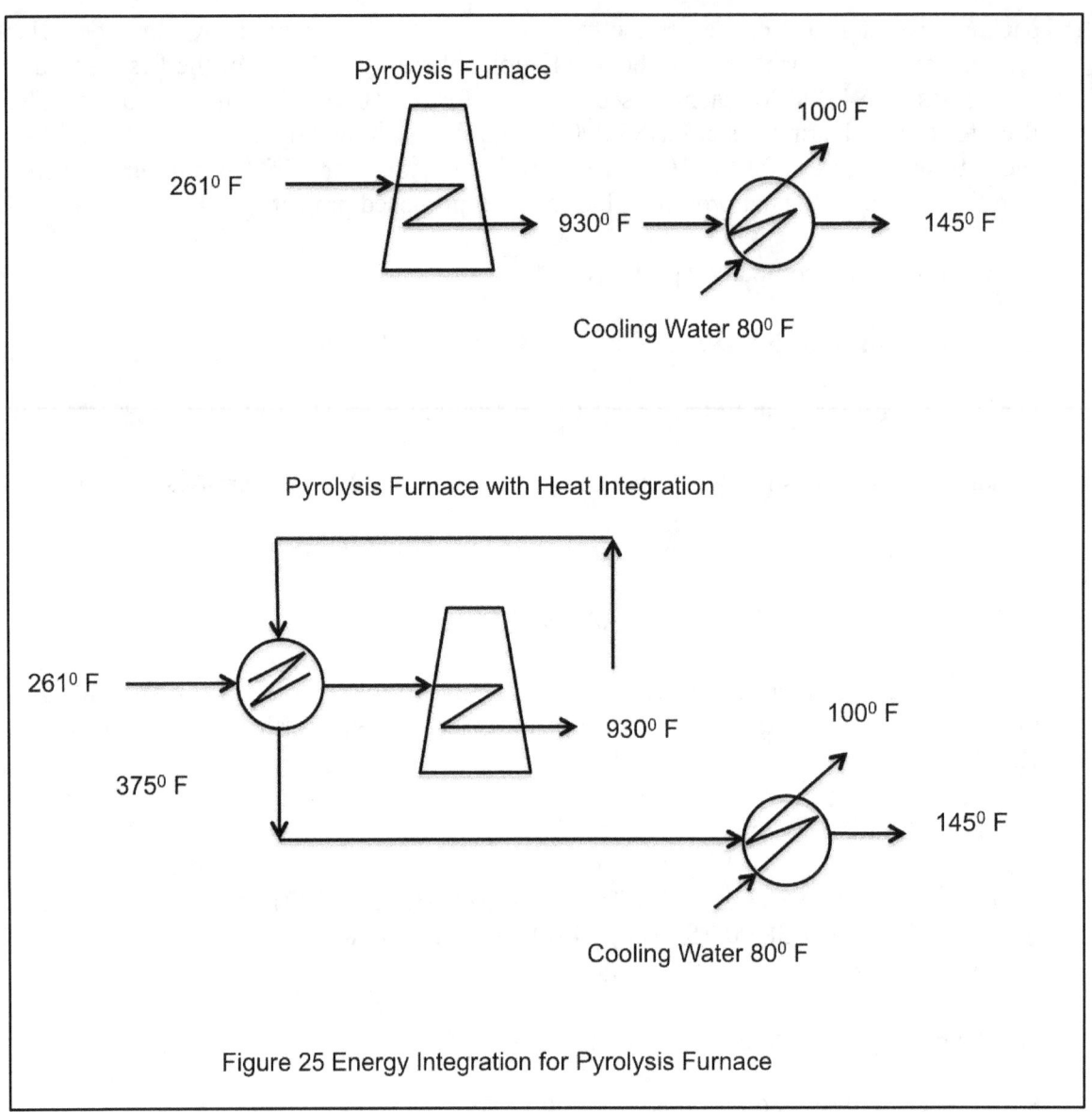

Figure 25 Energy Integration for Pyrolysis Furnace

The integration requires another heat exchanger with an installed cost of $150,000 and the energy savings are estimated to be $30,000 per year. The minimum attractive rate of return for the company is 10% before taxes, and the economic life is 10 years.

a. The net annual savings (loss) is from the difference between the energy savings and the equivalent uniform annual costs (EUAC). Converting the heat exchanger cost to an equivalent uniform annual costs gives:

EUAC = PW{i/[1 - (1 + i)$^{-n}$]} = $150,000{0.1/[1 - (1.1)$^{-10}$]} = $24,412 cost of capital

Net annual savings = $30,000 - $24,412 = $5,588

b. The net present value before taxes for the proposed energy integration, Equation 24.

$$NPV = -CF_o + A[1 - (1 + i)^{-n}]/i = -\$150,000 + (\$30,000/yr)[1 - (1.1)^{-10}]/0.1 = \$34,337$$

Based on these results, energy integration would be profitable. However, plant start-up must be considered in energy integration opportunities. For this case, additional heat would have to be provided on start-up until the furnace output temperature reached 930^0 F.

Rate of Return: The rate of return (ROR) is the interest rate where the net present value is zero, i.e. from Equation 23.

$$0 = -CF_o + \sum_{j=1}^{n} CF_j(1+i)^{-j} \tag{25}$$

To determine this interest rate, it is usually necessary to interpolate between two known values of the net present value. In Example 5 the net present value was 9.64 at an interest rate of 15% and -4.34 at 25%. Interpolating gives the rate of return for this case to be 21.9% for a zero net present value. Interpolating is preferred to solving an nth order polynomial. Multiple values of the interest rate can occur when there are positive and negative cash flows. This is the result of locating the roots of a polynomial. The following example gives an additional illustration of the evaluation of the rate of return.

Example 11. Evaluating the Rate of Return

A division of a company has been allocated $100,000 to invest at the start of the next fiscal year in cost-reduction projects. Three projects are under consideration and are summarized below.

Project	Estimated Economic Investment Required	Life (years)	Net Annual Cash Flow
A	$50,000	9	$16,600
B	$50,000	8	$15,000
C	$100,000	6	$30,000

The minimum attractive rate of return for the company is 20% for projects with these economic lives. Would a recommendation based on the rate of return for these projects be: 1. Invest in A only, 2. Invest in B only, 3. Invest in A and B, 4. Invest in C only, or 5. Seek other alternatives.

Alternatives are evaluated for investing $50,000 and $100,000 by comparing the rate of return for the projects with the minimum attractive rate of return of 20%. The rate of return, i, is the interest rate where the net present value is zero. For a uniform net annual cash flow, A, the equation for the net present value (NPV), Equation 24, is:

$$NPV = CF_o + A(P/A,i\%,n)$$

where CF_o is the capital investment and $(P/A,i\%,n) = [1 - (1+i)^{-n}]/i$ is the uniform series capital recovery factor.

For Project A: $0 = -50,000 + 16,600(P/A,i\%,9)$ or $(P/A, i\%,9) = 3.012$

Using Equation 18 with $(P/A, i\%, 9) = 3.012$ and $n = 9$ years, gives $i = 30.0\%$.

For Project B: $0 = -50,000 + 15,000(P/A,i\%,8)$ or $(P/A,i\%,8) = 3.3333$

Using Equation 18 with $(P/A, i\%, 9) = 3.3333$ and $n = 9$ years, gives $i = 25.0\%$.

For Project C: $0 = -100,000 + 30,000(P/A,i\%,6)$ or $(P/A,i\%,6) = 3.3333$

Using Equation 18 with $(P/A, i\%, 9) = 3.3333$ and $n = 9$ years, gives $i = 20.0\%$.

Summary: Project Rate of Return
          A   30.0%
          B   25.0%
          C   20.0%

The investment decision is to select Projects A and B because their rate of return is greater than the minimum attractive rate of return; and all of the available capital is used.

Recommended Application of Net Present Value, Rate of Return and Equivalent Uniform Annual Cost Methods: In the design of a new process the rate of return method is used to compare independent alternatives, i.e. company projects with other types of investments such as acquisitions, bonds, and certificates of deposit.   It is a standard measure used in capital budgeting.

Net present value is used to choose among alternatives. The net present value of projects can be added for a total net present value, while rates of return are not additive. Economic analyses used by companies employ more than one method, and they have developed experience on the significance of each of these methods when considering building new plants.

The equivalent uniform annual cost method is used for evaluating retrofits and debottlenecking in existing plants. This method predicts a net annual savings in operating cost for the company. This savings will end up as an annual increase in the company's profits. Each proposed retrofit can be compared on the potential savings for the plant.

Payback Period: This is the time required to recover the capital investment from the net profit, but it neglects the time value of money. The equation for the payback period is (Sepulveda, et. al., 1984):

$$CF_o = \sum_{j=1}^{PBP} CF_j \tag{26}$$

and if the yearly net profits A are uniform, then the payback period (PBP) is given by the following equation.

$$PBP = CF_o / A \qquad (27)$$

The pay back period is a simple and popular calculation, but it ignores the time value of money and should not be used for making economic decisions for that reason. It is sometimes called the *payout time*. However, there is a modification to the payout time called the **discounted payout time** that computes the number of years to have the cumulative discounted cash flows sum to zero, and this does include the time value of money (Ward, 1986).

To illustrate the calculation of the payback period using Example 5, the capital investment was to be $71.0 million, and the sum of the cash flows was $114 million for the five years. The payback period is determined by:

$$71.0 = 32.0 + 28.0 + 22.0(0.5) = 71$$

which gives a payback period of about 2.5 years. Although this time to recover the capital investment may sound attractive, it is not a good investment if the cost of money is at an interest rate of 25% as shown in the example.

Economic Price: The economic price is computed using the total product cost and the annual capital cost based on the rate of return required by the company as given by Equation 22 for EUAC.

Economic price =   [(total product cost , C_T +
                annual cost of capital, EUAC +
                annual capital expenditure, C_{cap})]/ product rate, m   (28)

Referring to Example 1, the economic price computed for the aniline process example is:

($46.3 million/yr + $2.37 million/yr + $0.5 million)/100 millions pounds/yr = $0.49/lb

It can be shown that the net present value is zero in the economic price evaluation, and the interest rate is the rate of return given by Equation 25. See problem 21.

Example 12 Economic Analysis of a Green Process for Acetic Acid Production

A "green" process is proposed to use cellulose waste in an anaerobic bioreactor to produce a 60% methane and 40% carbon dioxide gas product that is sent to a catalytic reactor to produce acetic acid. The process flow diagram is shown below in Figure 26 followed by the capital costs, product costs and revenue.

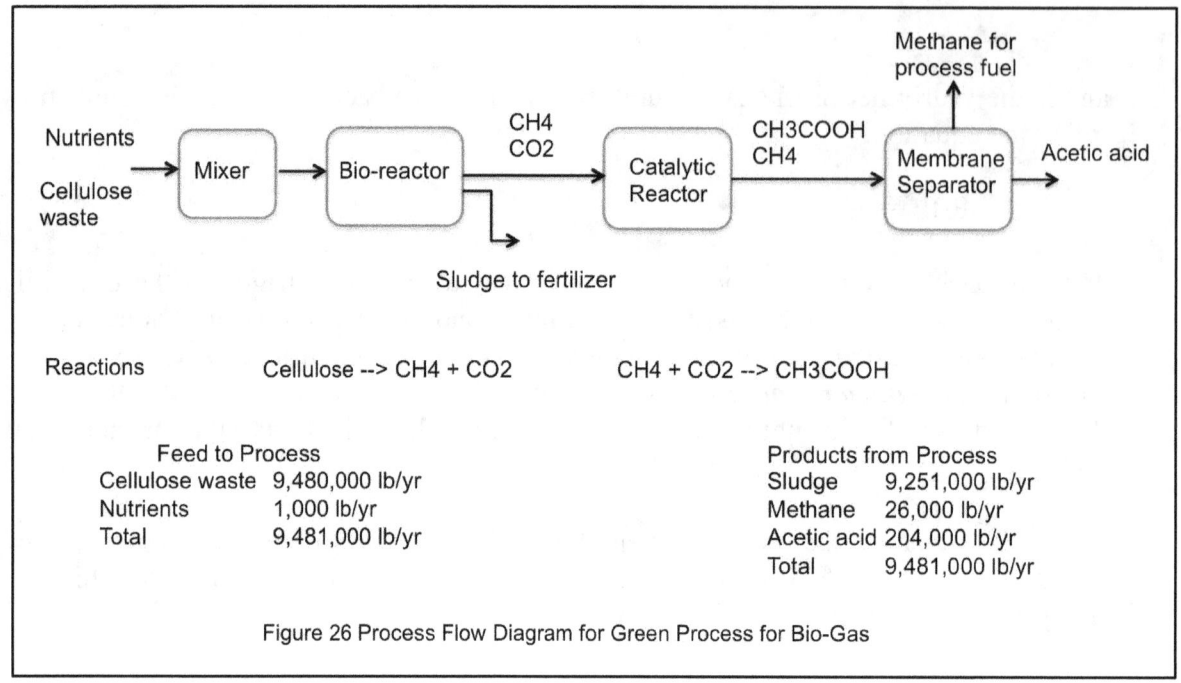

Figure 26 Process Flow Diagram for Green Process for Bio-Gas

Capital Costs

installed equipment cost $185,000

total capital cost $437,000

Product Costs

manufacturing cost $17,000/yr

general expenses $1,000/yr

total product cost $18,000/yr

Revenue

acetic acid $173,000/yr

sludge for fertilizer $8,000/yr

total sales revenue $181,000/yr

a. Computing the net annual cash flow before taxes. There are no annual capital expenditures.

$$CF = sales - total\ product\ costs - annual\ capital\ expenditures$$
$$= \$181,000 - \$18,000 - 0 = \$163,000$$

b. Computing the annual depreciation using the straight line method based on the installed equipment cost and no salvage value for a 10-year economic life.

$$D = \$185,000/10\ yrs = \$18,500/yr$$

c. Computing the annual net profit after taxes for a corporate tax rate of 35%.

Taxable income = net income - depreciation = $163,000 - $18,500 = $144,500

Taxes = (0.35)($144,500) = $50,575

Net profit after taxes = $163,000 - $50,575 = $112,425

d. Computing the rate of return for the proposed plant (net present value = 0) for a 10-year economic life.

$$NPV = 0 = CF_o + A[1 - (1+i)^{-n}]/i$$

where i is the rate of return to be determined for n = 10 years

$0 = -\$437,000 + (\$112,425)[1 - (1+i)^{-10}]/i$, solving by trial and error gives $i = 22.4\%$

Example 13 Economic Analysis for Aniline Process

The economic analysis for the aniline process is shown in Figure 27. It uses the information in Examples 3 and 4. These values were obtained using an Excel spreadsheet. The equations for the net present value, economic price and other relations given in the chapter were used to obtain these results. The net present value was $47.4 million, and the economic price was $ 0.43/ lb. Note that the copy of the spreadsheet in Figure 27 has up to eight significant figures for some of the numbers, but this is spreadsheet computation. The precision of the values is at best three significant figures.

Aniline Plant Net Present Value Analysis

Plant capacity (million pounds per year)		100
Plant installed cost or fixed capital investment, FCI	$	5,810,342
Total plant cost or total capital investiment, TCI	$	8,715,512
Total product cost	$	42,649,187
Annual expenditure for worn out equipment	$	500,000
Estimated annual sales	$	60,312,000
Economic life, years		10
Tax rate, 35%		0.35
Minimum attractive rate of return, 15%		0.15
Depreciation, straight line with no salvage value		
Net annual income before taxes = sales - total product cost	$	17,662,813
Net annual cash flow before taxes =	$	17,162,813
net annual income - annual expenditures for equipment		
Depreciation = plant installed cost/ economic life	$	581,034
Taxable Income = net income before taxes - depreciation	$	17,081,779
Taxes rate 0.35	$	5,978,623
Net income after taxes (before -taxes)	$	11,684,191
Net annual cash flow after taxes =	$	11,184,191
net annual income - annual expenditures for equipment		
Net Present Value =	$	47,415,352
-TCI + net annual cash flow *[1 - (1 + i)exp-n]/I		
Economic price	$	0.43
sales/prod rate = {TCI/[1 + (1.2)exp-10]/0.2] +total prod cost}/product rate		
Product flow rate (lb/yr) 12,565		

Figure 27 Economic Analysis for the Aniline Process

Benefit-Cost Analysis: The benefit cost ratio (BCR) is used in municipal and government projects, and it is defined by the following equation (Sepulveda, et. al., 1984).

$$BCR = (B - D)/C \qquad\qquad (29)$$

The benefit-cost ratio is the difference between the benefits, B, and the disbenefits, D, divided by the costs, C. For example, in a project to build a hydroelectric dam, the benefits would be electric power generation and possibly flood protection; and the disbenefits would be loss of productive farmland. The costs would include the construction and maintenance of the dam.

Although, benefits and costs can be estimated without too much difficulty, it is usually difficult to measure the disbenefits from the loss of wildlife habitat, scenic rivers and land loss in coastal marshes from lack of replenishing sediments. For projects to be considered by the U.S. Army Corp of Engineers, BCR should be about 2.0 or larger. The benefit-cost ratio is frequently incorrectly referred to as the cost-benefit ratio.

Equivalence: In comparing investment opportunities, there are times when two or more plans give the same result, the same net present value, for example, even though plans involve different interest rates and times. A plan that has a uniform annual series value of $8,000 for 25 years at an interest rate of 6% has a present value of $102,264. Another plan that has equal quarterly deposits of $156.07 for 40 years at a nominal interest rate of 6.0% has the same net present value of $102,264. These plans are said to be equivalent since both accumulated a present value of $102,264. This illustrates the concept of *equivalence*, i.e. being of equal value. The interest factors presented previously are called equivalence factors, also (Sepulveda, et. al., 1984). Other simple examples of equivalence include the following.

A uniform annual payment of $655.56 for 20 years at an interest rate of 8% is equivalent to a single payment of $30,000 in year 20.

A single payment today of $30,000 would be equivalent to an annual payment of $6,436.45 over 20 years at an 8% interest rate.

Investing $6436.45 for 20 years at a 15% rate is equivalent to $105,342 today. A uniform series of payments of $1,028 for 20 years at a 15% rate is also equivalent to $105,342.

The first two examples show that, at 8%, $30,000 in 20 years is equal to $655.56 a year for 20 years, which is also equivalent to $6436.45 in the present. The third example shows that investing two different amounts of money are equivalent to $105,342.40 today.

Sources and Cost of Capital

A company requires capital for construction of plants and their improvements. There are several sources, and each has its cost. *Equity capital* is supplied by the owners and shareholders, and dividends are paid for this capital as a share of the profits. *Borrowed capital* is supplied from lending institutions. Interest must be paid, and the capital repaid, and there is no profit sharing, DeGarmo, et.al, 1988.

Retained earnings from profits can be used as a source of capital for reinvestment, but stockholders expect 50% of profits to be paid as dividends.

Depreciation is a source of funds, also. If equipment is depreciated fully, then this tax deduction is lost. Consequently new equipment can be added to retain funds that would be lost to taxes.

A corporation acquires equity capital through the sale of stock, and the stockholders are the owners of the corporation. Common stock can be issued, and it does not guarantee a dividend. Preferred stock can be issued with a definite dividend guaranteed. Losses are limited to the value of the stock purchased, and stockholders are entitled to a share of the profits but are not liable for the debits of the corporation. Selling more stock to raise capital dilutes the value of the current stockholders investments. Also, there is double taxation where the corporation pays taxes on profits, and the stockholders pay taxes on dividends that are distributed by the corporation. Large companies are usually corporations. However, companies can be individual ownership or a partnership, and the owner or partners are taxed on the income from the company that avoids double taxation.

Borrowed or debt capital is used to finance projects such as new plants and expansions of existing plants, also. Borrowing funds for capital expenditures over a definite period of time is an advantage to stockholders. Interest is a business expense and is tax deductible. Sources of these funds include short-term notes from a lending agency such as a bank or insurance company. Bonds can be used, and they are long-term notes.

Income tax as a business expense reduces the effective cost of borrowed capital and hence the term "after tax weighted cost of capital." Most capital is obtained from equity sources, stocks. Smaller amounts are borrowed. Dividends are paid on equity capital, and interest is paid on borrowed capital. Lenders expect that no more than 50% of a project requiring capital expenditure will be borrowed capital to keep their risk at a reasonable level.

Example 14. A brief comparison of cost of borrowed and equity capital.

Borrowed capital: Consider $10,000 borrowed at 8%. The annual simple interest is $800.
  Net annual income before taxes is reduced by $800.
  Savings on income taxes at a tax rate of 35% is 0.35x$800 or $280
  Net after tax cost of interest is $800 - $280 = $520
  After tax interest rate is ($520/$10,000)x100 or 5.2%
  Debt capital only needs to generate 5.2% of added income to cover the interest to be paid on this capital.

Equity capital: Consider $10,000 in equity capital with a 8% dividend.
  Dividend is $800, and dividends are not tax deductible for the corporation or for the

stockholders.

Equity capital would have to generate a pretax amount of $800/(1 - 0.35) or
   $1,231 to pay dividends by the corporation.

Taxes to be paid on earnings are $1,231x0.35 or $431, leaving $800 to go to
   stockholders for dividends.

In summary, borrowed capital is less costly than equity capital for financing a project. However, lenders usually require about 50% equity capital be used for a project, except in the case of leveraged buy-outs. See the discussion in Garrett, 1989 about the Sterling Group, Inc. forming Cain Chemical, Inc. to operate seven leveraged buy-out petrochemical plants resulting in a billion dollar company with 1,500 employees.

Opportunity Costs: An opportunity cost (profit) is the funds that are not received if a particular project is rejected or not pursued. It is the amount lost if an opportunity is not taken. For example, a company considers upgrading a computer system, and it will receive a trade-in allowance. If the computer system is not upgraded, the trade-in allowance is not obtained. The amount of the trade-in allowance is an opportunity cost that is included in project evaluations for capital investment.

In considering replacement of equipment, the existing equipment will have a salvage value. If a replacement is purchased, the salvage value would reduced the purchase cost of the replacement. Consequently, salvage value is treated as an opportunity cost that would be incurred if the existing equipment were not replaced.

Taxes and Depreciation

In any project analysis, it is important to distinguish between before tax cash flows and after tax cash flows (BTCF and ATCF). (See Table 1.) In general, best project decisions are based on after tax cash flows, but before tax figures can often be used in the preliminary stages of analysis as an approximation for the desirability of a project. Under the current tax code, a corporation's first $50,000 in profits is tax exempt, and the rate is scaled from 15% to 39%, as shown in Table 8. Referring to the table, most corporations are in the top three brackets in taxable income. Consequently, in economy studies, the corporation is operating and making decisions where marginal profits are taxed at the highest rate (currently about 35%). Although tax rates are constant for marginal profits, after tax analysis is more difficult than before tax analysis because of the evaluation of depreciation.

Depreciation as defined by the U. S. Internal Revenue Service (IRS) is an annual tax deduction allowed to recover the cost for wear and tear, deterioration or obsolescence of property. Property can be tangible (except land) such as buildings, machinery, vehicles, furniture and equipment. Certain intangible property can be depreciated such as patents, copyrights and computer software (http://www.irs.gov/Businesses/A-Brief-Overview-of-Depreciation).

Table 8 Corporate Income Tax Rates for 2012 (2012 Instructions for Form 1120, U. S. Corporation Income Tax Return, http://www.irs.gov/pub/irs-pdf/i1120.pdf

Taxable Income

over but not over	Pay	plus % on Excess	of the amount over
$0 - $50,000	$0	15%	$0
$50,000 - $75,000	$7,500	25	$50,000
$75,000- 100,000	$13,750	34	$75,000
$100,000- 335,000	$22,250	39	$100,000
$335.000 - 10 million	$113,900	34	$335,000
$10 million - 15 million	$3.4 million	35	$10 million
$15million - 18,333,333	$5,150,000	38	$15 million
$18,333,333 - ...	$6,416,667	35	$18,333,333

Depreciation is based on capital. Capital goods are those accumulated in order to produce other goods. Two kinds of capital can be distinguished, fixed and working. *Fixed capital* is that which cannot be readily converted into a different sort of asset. It is "financially immobile" (Ulrich, 1984). A great majority of capital will be fixed. *Working capital* is the investment that puts the plant into production, or "money invested before there is a product to sell" (Ulrich, 1984). It can be estimated to be a tenth to a fifth of fixed capital, or as a value of one month's worth of raw materials.

Total capital (fixed plus working) does not total the expenses of operation. There are other indirect expenses such as overhead, taxes, insurance and *depreciation.* The amount of depreciation in any given year depends on the amount of *fixed* capital only. Income taxes are a function of depreciation.

Depreciation exists because anything valuable will lose some or all of that value with time. Holland, et al. 1983, has given four definitions of depreciation, all of which are correct. Depreciation can be seen as:

1. a tax allowance
2. a cost of operation
3. a means of building up a fund to finance plant replacement
4. a measure of falling value

The last definition is the most literal use of the term, and it is the least useful one in describing depreciation's importance in process economics. It is true that, when a cost entry for depreciation is made, it is hypothetically meant to represent an actual fall in value (which is, in reality, impossible to access accurately).

The accounting of depreciation is also a way to plan for replacing equipment, except actual replacement of obsolete equipment is rare. New technology and new methods of production will usually require totally new equipment.

Depreciation appears on a financial statement as a cost of operation, but this is equally misleading. Capital costs are depreciated, and in that way are translated into yearly operating costs, but the actual disbursement was made at the time of equipment purchase. The reason for depreciating the amount over time is this: capital equipment is an asset to the plant; each year, the value of that asset declines and even though there is no out of pocket cost, the declining value is a real cost to the plant. However, since capital costs are incurred at the inception, depreciation costs are calculated yearly only for the calculation of taxes.

Depreciation as a tax allowance is the most important consideration when planning a project. The depreciation entry in the cost column serves to lessen taxable income, and it can make the crucial difference in profitability. This entry may have little to do with the actual physical depreciation of the asset, for tax laws have increasingly standardized the ways in which depreciation is calculated, thus making the depreciation entry less a measure of an individual piece of equipment's useful life. Still, the expense should be viewed as a measure of useful life.

To determine the amount of depreciation for tax purposes, a few values need to be known. The *depreciable base* consists of fixed capital only. Land, however, is exempted from the depreciable base, because it is thought always to retain all of its original value, and so cannot be considered a depreciable item. The *write-off life* is the hypothetical life of the asset. The IRS has guidelines for write-off life of equipment, and plant equipment is depreciable over about ten years. Computers have a write-off life to about 5 years. Next various methods of calculating depreciation will be explained.

Basically, there are four methods of computing depreciation: straight line, statutory, accelerated, and decelerated. Straight line assumes a constant loss of value over time, while accelerated depreciates more in the early years of the asset and decelerated depreciates more in later years. In the statutory method, loss in value is accelerated and specified in the IRS tax code. The straight-line method is used frequently since it is straightforward and generally a good approximation. The statutory method (U. S. Master Tax Guide, 2013) is the Modified Accelerated Cost Recovery System (MACRS), and according to the U. S. IRS it is the "proper depreciation method for most properties." Additional details are given in IRS publication, Publication 946, How to Depreciate Property, (http://www.irs.gov/publications/p946/index.html).

Straight-Line Method: With this method the annual depreciation, D_r, is constant. To evaluate the depreciation charge for the rth year, it is computed as the difference between the original cost P and the estimated salvage value S divided by the estimated service life n in years, i.e.

$$D_r = (P - S)/n \qquad\qquad (30)$$

Then the book value at the end of the rth year, BV_r is obtained my subtracting r times the depreciation D_r obtained from Equation 30 as shown below.

$$BV_r = P - rD_r \tag{31}$$

For example, the original cost is \$20,000; and the service or write-off life is 12 years for a piece of equipment. With no salvage value, the depreciation for year 1 (r = 1) is: $D_1 =$ (20,000 - 0)/12 = \$1,667; and the book value is: $BV_1 = 20,000 - (1)(1,667) = $18,333.

Statutory Methods: The rules for depreciation computation were given in the Tax Reform Act of 1986. The U. S. Tax Code (U. S. Master Tax Guide, 2013) requires that equipment placed in service after 1980 must use the Accelerated Cost Recovery System (ACRS). After 1986 the Modified Accelerated Cost Recovery System (MACRS) must be used. Prior to 1981 equipment must be depreciated according to the method originally chosen, and the ACRS or MACRS cannot be used (http://www.irs.gov/publications/p946/index.html). The two major advantages of ACRS and MACRS depreciation methods are that computations are made on "property class lives" that are less than "actual useful lives" and that the initial cost is not reduced by the equipment's salvage value.

A table of class lives and recovery periods for a number of industries is given in Table B-2 of IRS publication, Publication 946, How to Depreciate Property (http://www.irs.gov/pub/irs-pdf/p946.pdf). For petroleum refining from this table, the Asset Class is 13.3, the Class Life is 16 years, the Recovery Periods for General Depreciation System (GDS) is 10 years and the Alternative Depreciation System (ADS) is 16 years. For the manufacture of chemicals and allied products from this table, the Asset Class is 28.0, the Class Life is 9.5 years, the Recovery Periods for General Depreciation System (GDS) is 5 years and the Alternative Depreciation System (ADS) is 9.5 years.

Using the ACRS or MACRS, depreciation is calculated by multiplying the equipment cost, C, by a factor, i.e., $D_j = C \bullet$(factor) where the factor is taken from one of a number of tables depending on the equipment. The first step is to determine the property class of the asset. Each depreciated asset is placed in a MACRS GDS property class. There are six property classes: 3-year property class (e.g. special manufacturing tools), 5-year (e.g. research equipment), 7-year (e.g. office equipment), 10-year (e.g. petroleum refining equipment), 15-year (e.g. sewage treatment plant), 20-year property (e.g. utility services for electricity, gas, water, steam). Then the depreciation schedule is read from a table. Depreciation for the 3-, 5-, 7- and 10- year property classes are based on ***double declining balance depreciation*** with conversions to ***straight line depreciation*** in the fourth through the ninth year depending on the class of the property to maximize the deduction. This method is described in detail by Newnan, et al., 2004 and in the *U. S. Master Tax Guide, 2013* with the necessary tables to perform the calculations. These tables are too lengthy to include here.

Double-Rate Declining Balance Method (Grant, et. al, 1982): With this method the depreciation is computed at double the straight line depreciation rate, and it is one way to do the declining balance method. If f is the double-declining balance rate given by:

$$f = 2.0/n \tag{32}$$

then the book value is computed by the following equation.

$$BV_r = P(1 - f)^r \qquad (33)$$

The depreciation at year r is given by the following equation.

$$D_r = (BV_{r-1} - BV_r) = fBV_{r-1} \qquad (34)$$

Using information from the straight-line method, the double-declining balance rate is $f = 0.2/12 = 0.167$ by Equation 32. Then the book value for year 1 ($r = 1$) is: $BV_1 = (20,000)(1-0.167) = \$16,660$, and the depreciation is: $D_1 = (20,000-16,660) = \$3,340$.

In this method the net salvage value is not evaluated as part of the procedure. The method can have the book value become smaller than the net salvage value. However, IRS regulations require that the book value not be less than the salvage value, and the depreciation must stop at this point (Sepulveda, et al., 1984).

Sum-of-Years-Digits Method Sepulveda, et al. (1984): This is an accelerated method using the sum-of-year-digits, SY, that depreciates about 75% of the cost in the first half of the service life. SY is given by the following equation.

$$SY = \sum j = 1 + 2 + ... + n = n(n+1)/2 \qquad (35)$$

where n is the service life. The amount of depreciation D_r is computed as follows:

$$C = (P - S)/SY \qquad (36)$$
and
$$D_r = (n+1-r) C \qquad (37)$$

Then the book value is obtained from the following equation.

$$BV_r = P - C[SY - (n-r)(n-r+1)/2] \qquad (38)$$

Using the information from the straight-line method, the sum-of-year-digits is: $SY = 12(12+1)/2 = 78$, and the value of $C = (20,000 - 0)/78 = \$256$. Evaluating Equation 37: $D_1 = (12+1-1)256 = \$3,072$, and the book value from Equation 38 is: $BV_1 = 20,000 - 256[78-(12-1)(12-1+1)/2] = \$16,904$.

Sinking Fund Method (Sepulveda, et al., 1984): This method depreciates equipment with an imaginary sinking fund that is equivalent to the company making a series of equal annual deposits to have an amount equal to the cost of replacing the equipment at the end of its service life. The amount of depreciation in any given year is equal to the amount of the sinking fund plus interest. This method is used when the replacement of the equipment is assumed to cost the same as the original. It is the only method in which depreciation increases in time. The accumulated depreciation in year r, AD_r, is given by the following equation.

$$AD_r = (P - S) (A/F, i\%, n) (F/A, i\%, r) \tag{39}$$

The depreciation for year r is computed by difference, i.e.:

$$D_r = AD_r - AD_{r-1} \tag{40}$$

Then the equation to evaluate the book value is:

$$BV_r = P - (P - S) (A/F, i\%, n) (F/A, i\%, r) \tag{41}$$

Using the information from the straight-line method and an interest rate of 8%, the accumulated depreciation for the first year is: $AD_1 = (20,000 - 0) (0.0527) (1) = \$1054$ using Equation 39, and the depreciation is: $D_1 = 1054 - 0 = \$1054$ using Equation 40. Then the book value is $BV_r = 20,000 - 20,000 (0.0527) (1) = \$18,946$ from Equation 41.

In this method the annual depreciation increases geometrically with time. Tax laws had required that this method be used only with equipment that has to be replaced with equipment that cost at least as much as the original. This does not allow a company to take a total depreciation allowance that is greater than the equipment's current net adjusted cost. Consequently, this method is used very seldom.

Comparison: A comparison of four depreciation methods is given in Figure 28 from Sepulveda, et al. 1984 and three methods in Figure 29 from Soares, J. B., et al, 2006. In Figure 28 the annual amount of depreciation of $1,000 is given as a function of time in years. In Figure 29 the ratio of the book value at year t to the depreciable capital investment is given as a function of time in years. As can be seen in the figures, the straight-line method gives a constant annual depreciation amount over the economic life. The others have the annual amount vary over the economic life. The statutory methods, ACRS and MACRS, use a combination of double declining balance and straight line depreciation. Prior to 1981 the method would be chosen that minimizes taxes paid by the corporation. If profits were higher in the initial years, the double declining balance or sum-of-digits methods would be preferred. If profits were higher in the later years of the economic life, the sinking fund method would be used. However, the sinking fund method is seldom used.

Capitalized Assets versus Expenses: High expenses reduce profits that in turn reduce income tax. Consequently, the minimum tax would be obtained if all expenditures, capital, operating, etc. were an expense. However, the tax code allows for depreciation of equipment, and there are tax code regulations about what can be an expense and what is capital equipment. Most companies use a cut off value for expenses, $500 for example, and above this figure items are capitalized. This procedure avoids keeping track of small pumps, valves, instruments, etc. Also, when an item is capitalized, it is placed in a group or asset class with other similar items.

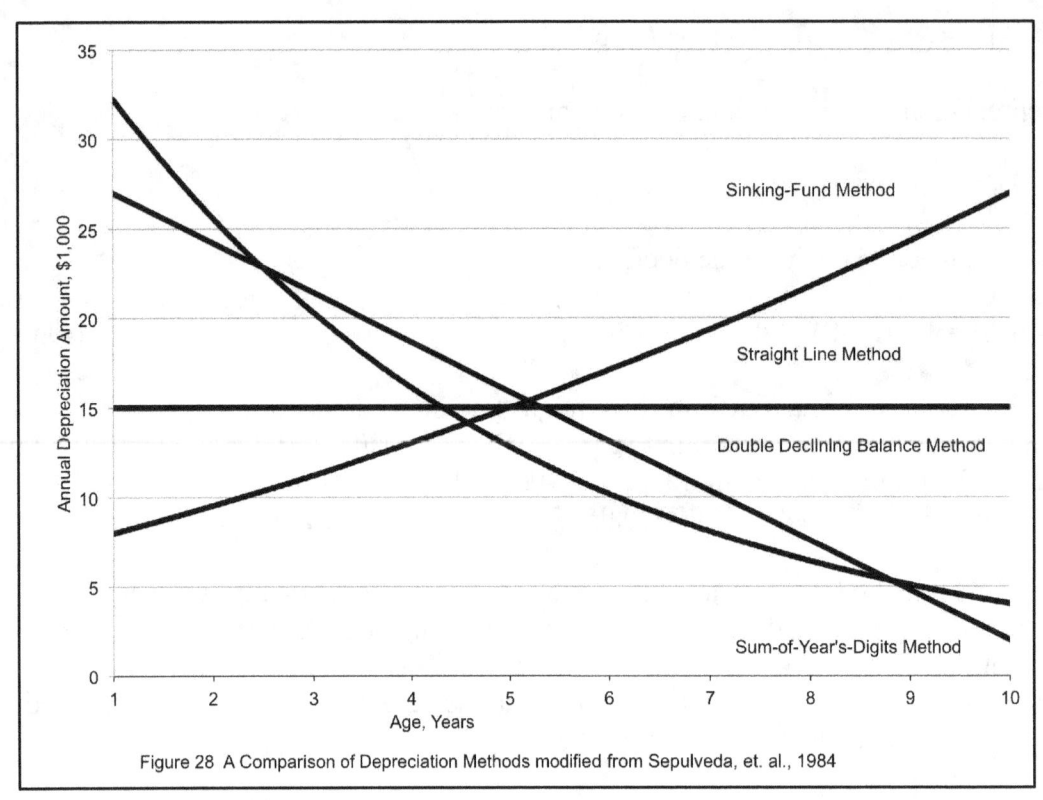

Figure 28 A Comparison of Depreciation Methods modified from Sepulveda, et. al., 1984

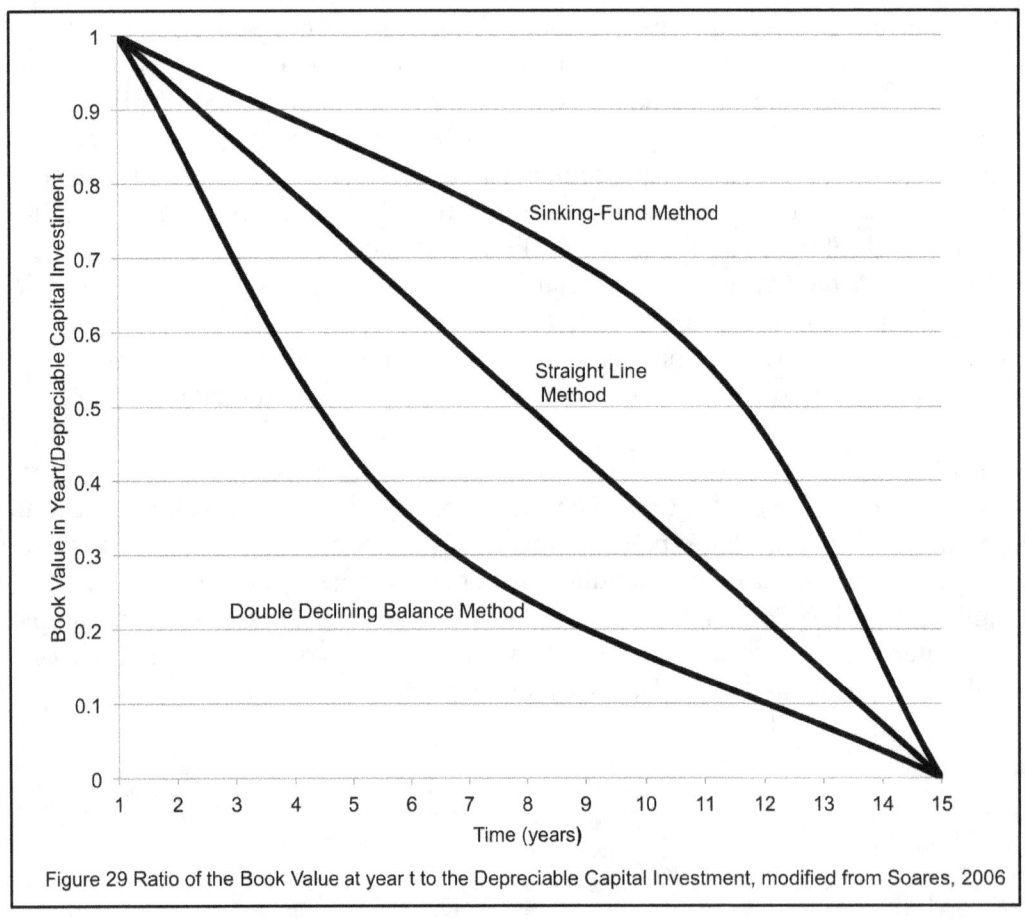

Figure 29 Ratio of the Book Value at year t to the Depreciable Capital Investment, modified from Soares, 2006

Group Accounts: Often it will be more convenient to group assets that are bought in the same year for depreciation purposes. Items that have the same useful life can be pooled into a *group account*, regardless of whether they were bought in the beginning or the end of the year. A *classified account* groups items according to use, regardless of useful life, and a *composite account* includes items that have diverse lives and uses. With classified or composite accounts, the depreciation rate for any year is found by determining depreciation for each item or each group of similar items, and that total is divided by total cost. Any of the above depreciation methods will work.

Comparison and Summary: Comparing the depreciation methods, the accelerated methods are those for which the accumulated depreciation exceeds the straight-line method. However, the sinking fund curve is below the straight-line, meaning it is decelerated. The amount of depreciation is ultimately the same for all, except in double declining balance, in which the depreciation total approaches, but never reaches, the full amount. In instances where the salvage value is significant, the double-declining rate's acceleration effect is enhanced, since salvage value is not considered in calculation.

Advantage of Accelerated Depreciation: It might seem that, since the depreciation amount is ultimately the same, acceleration is simply a shortsighted way of grabbing tax credits in the shortest time possible. Not at all. The advantage to accelerated depreciation is very real. With money's value in time, the declining balance and sum of year's digits methods, because they are accelerated, yield a higher present worth for net income after taxes. The discount for the tax savings in earlier years is not as great as the discount in later years.

Sometimes, especially with an accelerated depreciation method, the depreciation charge exceeds the actual before tax cash flow in a given year. When this occurs, realize that, although each investment decision is treated separately in the economic study, the income generated and taxes paid are part of the cash flow for the entire firm. Hence, if the depreciation charge is $1,000 and income is $600 for that year, then the other $400 of depreciation is not lost, but is subtracted from the general before tax cash flow of the firm. For purposes of analysis, this extra benefit to the firm can be included in the economy study of the specific investment decision in the following manner, which shows an after tax profit greater than the actual before tax receipts.

Before Tax Cash Flow, BTCF	$600
Depreciation	1000
Taxable income	- $400
Tax at 33%	- $132
After tax cash flow, ATCF	= $600 + $132 = $ 732

Every year, regulations on depreciation are being revised, and less choice is given to the corporation (or individual) in how they may claim depreciation deductions. The methods outlined above are important to know, since next year or thereafter they may be the basis of the newest law. Since 1981, however, formulas for depreciation have been set by the Accelerated Cost Recovery System (ACRS). With the ACRS, all assets can be grouped into one of four categories, each with a corresponding recovery period, and depreciation percentages are set for each year. This greatly simplifies matters, being very amenable to *group asset* depreciation.

Note that equipment is not subject to depreciation prior to the year it has been placed in service, that is, when it is not yet ready for use. A plant that takes 3 years to build is not depreciable until it is completed and placed in service. However, is makes no difference to the yearly deduction if the equipment was placed in service during the beginning of the year or in late December.

Example 15. Influence of Depreciation on Capital Investment

A company can reduce operating costs by $16,300 per year for 12 years by automating a process with a new digital control computer. The cost of the computer and control system is $78,000, and straight-line depreciation over 12 years is used with zero salvage value. If the applicable income tax rate is 38%, determine the rate of return after taxes.

For the new control system:
Operating cost is reduced by $16,300 per year for 12 years or an annual increase in profit.
Equipment cost is $78,000
Straight-line depreciation for 12 years with no salvage value
Income tax rate is 38%

The net profit is reduced by depreciation for tax purposes, and the change in the net profit per year before taxes from depreciation = $16,300 - ($78,000 - 0)/12   = $9,800

Taxes on the depreciated net profit = $9,800x0.38 = $3,724

The equation for the net present value (NPV) for uniform annual cash flow (A) is:

$$NPV = CF_o + A(P/A,i\%,n)$$

where CF_o is $78,000, the capital investment for the control system and A is the change in net annual income after taxes = $16,300 - $3,724 = $12,576

Computing the rate of return (interest rate at zero net present value).

$$0 = -78,000 + 12,576(P/A,i\%,12) \quad and \quad (P/A,i\%,12) = 6.202$$

Using Equation 18, P/A = 6.202 at 12 years, the above equation gives i = 12.0%, which is the rate of return after taxes.

Amortization, Depletion and Investment Tax Credit

Amortization is the recovery of capital expenditures that are not ordinarily deductible. The method is similar to depreciation and uses straight-line depreciation. The portion of the basis of the equipment that is recovered through amortization may not be depreciated. For example, pollution control equipment can be amortized over the first 15 years of its useful life. Also, intangibles can be amortized over a 15-year period beginning at the month of acquisition. Intangibles can include goodwill, workforce in place, patents, copyrights, formulas, designs, licenses, permits, trademarks and trade names. Details on amortization are given in the *U. S. Master Tax Guide, 2013*.

Depletion is a deduction allowed in determining the taxable income from natural resources. The deduction is similar to depreciation in that it allows the recovery of an asset over the resource's production life. Depletion is the exhaustion of natural resources such as mines, wells and timberland from production. Depletion can be based on amount or percentage produced per year. The percentage depletion deduction cannot exceed 50% of the taxable income from a property, except 100% for oil and gas production. Details on depletion are given in the *U. S. Master Tax Guide, 2013*.

Investment tax credit is a one-time credit against income taxes. It is added to the after tax present worth in an analysis. These tax credits are allowed by Federal and state governments for purchase of equipment, employment of various workers, and new plants, for example. Details on investment tax credits are given in the *U. S. Master Tax Guide, 2013*.

Economic Life Evaluation

Operations and maintenance cost increase as equipment ages. Also, the book value from depreciation decreases with time. At some point the sum of these two costs reaches a minimum and then increase. The age of equipment at this minimum point is the economic life of the equipment. This evaluation is illustrated with the following example (Lindeberg, 1998)

Example 16 Evaluation of the Economic Life of a Furnace

A plant has multiple high temperature furnaces that operate under sever conditions. The initial cost of a furnace is $120,000, and the maintenance cost and salvage value are given below.

Year	Maintenance Cost	Salvage Value
1	$35,000	$60,000
2	$38,000	$55,000
3	$43,000	$45,000
4	$50,000	$25,000
5	$65,000	$15,000

The equivalent uniform annual cost, EUAC, is evaluated for each year to determine the economic life. An interest rate of 8% is used.

Year 1
$$EUAC = 120,000(A/P,8\%,1) + 35,000(A/F,8\%,1) - 60,000(A/F,8\%,1)$$
$$= 120,000(1.08) + 35,000(1.00) - 60,000(1.00)$$
$$= \$104,600$$

Year 2
$$EUAC = [120,000 + 35,000(P/F,8\%,1)](A/P,8\%,2) + (38,000 - 55,000)(A/F,8\%,2)$$
$$= [120,000 + 35,000(0.9259)](0.5608) + (38,000 - 55,000)(0.4808)$$
$$= \$77,300$$

Year 3
$$EUAC = [120,000 + 35,000(P/F,8\%,1) + 38,000(P/F,8\%,2)](A/F,8\%,3) +$$
$$(43,000 - 45,000)(A/F,8\%,3)$$
$$= [120,000 + 35,000(0.9259) + 38,000(0.8573)](0.3880) +$$
$$(43,000 - 45,000)(0.3080)$$
$$= \$71,200$$

The evaluation is continued, and for year 4, the EUAC is $71,700. Consequently, the minimum occurs at year 3, and the furnace has an economic life of 3 years.

Sensitivity Analysis

Sensitivity analysis is an important method of quantifying the effect of uncertainty in the parameters used to evaluate the profitability of a project. It determines how "sensitive" a project is to some foreseeable change in the parameters that have been estimated, such as selling price, cost of materials, or tax rate. A sensitive project is one whose desirability is highly affected by a small change in any certain parameter. A project is insensitive to a parameter if wide ranges of values do not alter the conclusions of the study. Typically, a base case is used, and the present value or rate of return is evaluated for variations in parameters such as product price, sales volume, plant cost and size, working capital, etc. For example, a plant design is based on a capacity that will produce 4,000 items per unit time with a net present value of $3.0 million based on a projected selling price of $0.50 per unit. The sensitivity of the net present value to selling price of $0.25 and $0.75 per unit is -$0.5 million and $4.2 million. These sensitivity values can be obtained using a Monte Carlo simulation using statistical values for the parameters.

Differential price changes occur when income tax deductions for depreciation are not indexed for current dollars. If equipment costing $100,000 is depreciated over a period when inflation is 10%, the costs and receipts go up as well, but the deduction for tax purposes does not. Thus, the corporation is taxed on a higher percentage of income than it would be if there was no inflation, or if depreciation was indexed.

Risk Analysis

The sources of risk are uncertainties in variables that are estimated: disbursements, receipts, length of time the project will be in service, salvage value, and tax rate. However, a general rise in prices due to inflation is not a consideration, since receipts should change along with disbursements. However, differential price changes should be included in the analysis when individual prices vary in contrast to the general price level, if they can be foreseen. Differential prices may be estimated if revenues rely on contracts that are not indexed for inflation, or if governmental price controls apply. Fluctuations in oil and natural gas prices are other examples.

Sensitivity analysis is considered the first step in risk analysis, and it seeks to show the effects of changes in the key elements of the economic evaluation. For example, if the probability is high that the plant will not be used at capacity, then the desirability of the project is severely lessened. Sensitivity and probability analyses are used to estimate the elements of uncertainty.

Risk analysis attempts to quantitatively evaluate risks associated with research, development, economics, politics, natural disasters, and other possible uncertainties. Probability estimates for these risks can be obtained from past experience, simulations, expert estimation and experimental data, among others. The objective is to use risk-weighted expected values of the profit to maximize the income in selecting among projects accounting for risk associated with them. Risk-weighted expected values are the sum of the product of the probabilities and the associated profit for that outcome. The profit can be measured by the net present value, annual worth, or other appropriate economic values.

A simple risk assessment for economic decision analysis involves the following steps. For each project under consideration, determine the range of possible outcomes that would affect the profit, e.g. the range in product selling price. Evaluate the profit over this range of possible outcomes. Separately, estimate the probability of occurrences of the possible outcomes, e.g. the probability the selling price will be at the low value when the plant is constructed (a very unfavorable situation). Then evaluate the weighted average of the profit by computing the sum of the profit and associated probabilities. This weighted average profit is an estimate of the expected value of the profit. These expected values for the projects are used to rank them in order of economic potential with risk incorporated in the comparison.

A simple illustration of the separate determination of profit and probability is given in this example. Consider having the opportunity of paying for a chance to bet on the flip of a coin. Each toss cost a quarter, but the payoff is $1.00. The expected value of the game is the sum of the probabilities of the possible outcomes, both 0.5, times the payoff, either $1.00 or 0. Thus the expected value is (0.5 x $1.00 + 0.5 x 0) = $0.50. Even though, for any one toss, you may not receive $0.50; this is a bargain at a quarter a game. However, if the price of playing the game were exactly fifty cents, then the odds would not be for you or against you.

Determining the expected value of any given project consists of summing the product of the probabilities of all possible events times their corresponding profits. The following example illustrates this concept for an operating plant.

Example 16. Applying Expected Values from Weather Impacts for Risk Analysis

The supply of seafood is directly related to weather conditions at the fishing grounds. The percent of capacity that a seafood plant will be used in any one year was predicted using data from past years about what likely weather conditions and consequent catch. With this information, it was estimated that the probability of the plant being used at 75% capacity in any one year was 0.30, the probability for a full capacity year was 0.45, and the probability of usage at 125% of capacity was 0.25. Using this information, calculations of expected value of the annual worth are shown below.

Percent Capacity	Annual Worth	Probability	Probability•Profit
75%	$ 22,314	0.30	$6,694
100%	$124,045	0.45	$55,820
125%	$212,151	0.25	$53,037
		1.00	$115,552 Expected Value

Note that the sum of the probabilities is equal to one. This will always be the case. This example gives the expected value of the annual worth, but any of the other methods of project evaluation could have been used.

The following example illustrates another way risk analysis is used. Risk can be used to reduce the impact on any one investor.

Example 18. Applying Risk Analysis to Reduce Risk from an Investment

An oil company is considering four alternatives in drilling an exploratory well in a new region. One is to drill the well alone. The second is to have an equal partner. The third is to let a partner pay the drilling expense for 50% of the profit. The fourth is not to drill this well and use the funds elsewhere. The following table gives the net present value for each alternative in the event of success or failure.

Possible Outcomes	Drill Alone	Drill with Equal Partner	Partner Drills	Do Not Drill
Dry hole (failure)	-500	-250	0	0
Producer (success)	3000	1500	1500	0

For this example there are only two possible outcomes (success or failure). If the probability for success is a value P_s, then $(1 - P_s)$ is the probability of failure. Let NPV_s be the net present value for success, and NPV_f for failure. The following equations show that there is a linear relation between the probability of success and the expected value, EV.

$$EV = P_s\, NPV_s + (1 - P_s)\, NPV_f = P_s\, (NPV_s - NPV_f) + NPV_f$$

Substituting values for the net present values for the four cases gives the following equation:

1. $EV_{Alone} = \$3500\, P_s - \$500$
2. $EV_{Partner} = \$1750\, P_s - \$250$
3. $EV_{P.\,Drills} = \$1500\, P_s$
4. $EV_{Do\,Not\,Drill} = 0$

These linear equations are shown in Figure 30, and the EV_{Alone} is larger than the other two for values of P_s of greater than 0.25. A precise value for the probability of success may not be known, and these equations help bound the range of significant values. For any probability of success less than 0.25, having a partner drill would be the best alternative.

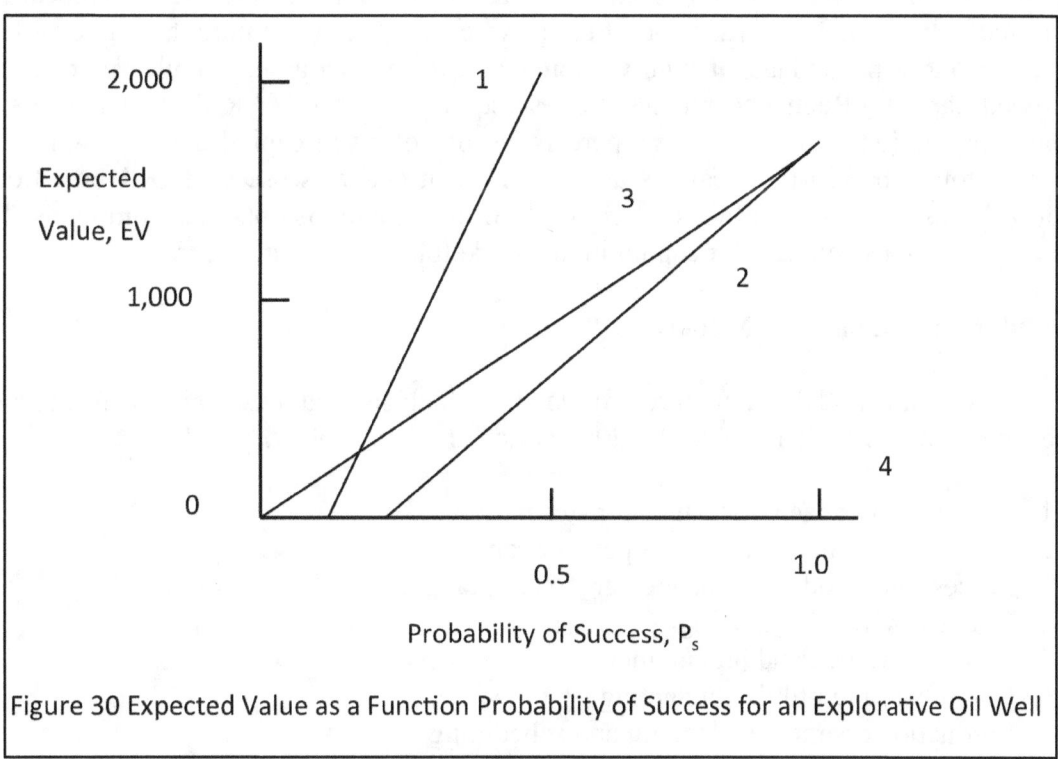

Figure 30 Expected Value as a Function Probability of Success for an Explorative Oil Well

There are times when the expected value is an inappropriate tool for decision-making. Expected values account for variations over a range of possible outcomes. If the immediate outcome is crucial, such that any risk is unpalatable, then the expected value is not a true measure of project worth. One common example is health insurance. People do not buy health insurance with the hope of making money. Rather, they are paying for the prevention of a financial disaster in case they are faced with a debilitating or long-term illness. The value of being sure about preventing financial disaster is more important to them than the risk of losing money in the transaction. That the expected value of health insurance is lower than the price paid for it is evidenced by the fact that insurance companies are profit-making businesses. To them, expected values and probability of illness are pertinent information; to the insurance policy holder, the knowledge that he is at risk is more important than knowing that he is probably losing money.

When all risks are unacceptable, expected values should not be used. If an investment of $5,000 yields a 8% return, or $400 a year, and a competing investment of $5,000 yields a 75% return, but with only a 15% likelihood of occurrence, the investors would have to choose between a sure $400, or an expected value of $562.50. The smart choice that is, the economically sound choice is the second investment, unless the chance of failure would mean disaster for the investor.

One Final Note about MARR

Company management determines the after-tax minimum attractive rate of return (MARR) based on funds available for investment and their source and cost. Also, projects that improve or expand operations and their associated risks determine the allocation of funds and help set the MARR. In addition, private corporations require the highest MARR, and use a large percentage of equity capital and a corresponding smaller percentage of borrowed capital. Regulated public utilities require after-tax MARR's that are less than private corporations, and use a large percentage of borrowed capital due to low risks and stable customer base. Municipalities and government agencies use a MARR based on the before-tax cost of borrowed funds. Almost all of their capital is obtained from loans. These three organizations compete for capital in the marketplace in varying degrees.

Formulation of Economic Models

Economic models are required for decision analysis, and these have been categorized as seven planning problems by Kelly, 2004 which are listed below.

1. Facilities or capacity expansion investment
2. Joint-venture allocation and apportionment
3. Process and product exchange agreement planning
4. Demand portfolio planning
5. Supply purchase and procurement
6. Chemicals and utilities budgeting
7. Production operations planning and scheduling

Economic models for the first three are for long range or long term planning that cover months to years. The next three are for near term planning that cover weeks to months. The last one on productions operations has a time frame of days to weeks. The time scale described here is illustrated in Figure 3. These economic models are applied to single periods of time e g., one week, for production scheduling and to multiple time periods, e g., five years, for production planning. Production scheduling is driven to meet orders for products (order-driven), and production planning is said to be forecast-driven. A planning horizon is used where the near-term horizon has frozen demands (orders) and the longer-term horizon has flexible demands (forecasts).

In economic models for planning, it is convenient to use a "value added" or "netback" model that is the difference between the gross profit and the cost of raw materials, utilities and other significant operating costs such as those associated with environmental emissions. Taxes are not included. The net present value is used for investment decisions, and there are five models that are used for operations and forecasting.

Economies of Scale: When there is a linear decrease in selling price as a function of quantity sold, the selling price can be expressed as:

$$\text{selling price} = b - a*\text{quantity} \tag{45}$$

and sales are then represented by selling price*quantity or for this case:

$$\text{sales} = \text{selling price}*\text{quantity} = b*\text{quantity} - a*(\text{quantity})^2 \tag{46}$$

In this case, the equation for sales is a quadratic function of quantity sold.

Fixed-Charge Capacity Expansion: This form of the economic model describes a discontinuity when extra capacity is required to meet demand but the existing plant capacity is exceeded. This leads to a decision to expend capital to build a new plant and the associated time lag to get this facility in production and to have an increase in gross profit. This economic model can be formed several ways as described by Kelly, 2004. These include creating an economic model with a binary variable for each fixed-charge capacity expansion decision, using a least squares fit to a fixed-charge capacity expansion term in the economic model, and use smoothing functions to approximate the step functions from binary variables. Kelly, 2004, provides more details.

Price Elasticity: Price elasticity, E, is the ratio of the percentage change in quantity, Q, to the percentage change in selling price. If the price elasticity is constant over a range of demand, then the following equation can be used for the selling price, Kelly, 2004.

$$\text{selling price} = a/Q^{1/E} \tag{47}$$

where a is defined as the price above which the quantity, Q is reduced to unity. Then sales in the economic model can be written as:

$$\text{sales} = \text{selling price}*\text{quantity} = a/Q^{1/E} * Q = a\, Q^{(1-1/E)} \tag{48}$$

Price elasticity is important in many petrochemical markets. The difficulty in including elasticity in the economic model is identifying the price above which the quantity is reduced to unity. More details are provided by Kelly, 2004 and in Appendix B on Supply, Demand and Price Elasticity.

Tiered Pricing or Quantity Discount Pricing: This method of determining the selling price is the discontinuous version of the economies of scale method, and uses a step function for the sales price according to the quantity sold. Sales in the economic model can be modeled with a step function as in the fixed-charge capacity expansion method or with peace-wise linear separable functions. A general equation for n steps in the quantity sold for the selling price is given by Kelly, 2004.

Premium Quality Pricing: This method adjusts the selling price incrementally above a target quality, $Quality_{Target}$, and is referred to as a performance based technique. Premiums are levied based on a key material or key quality. An example is the incremental price on gasoline, regular and two increments above regular. A way to represent the selling price for this case is:

$$selling\ price = b + a*(Quality - Quality_{Target}) \tag{49}$$

and sales is represented by the following equation:

$$sales = [b + a*(Quality - Quality_{Target})]* Quantity \tag{50}$$

In this case, the equation for sales is a linear function of the quantity sold.

Nonlinearities in Constraints: There are nonlinearities in constraints that are based on material and energy balances and the economic model. These are described in detail by Kelly, 2004 and are summarized here. First is called "quality dependencies," and these equations are approximations to the process model for large plants that relate the inputs to outputs and plant operating conditions. The second is called "weight based qualities," and these qualities have to be divided by the density of the material to have a "volume based quality" that is required for blending laws. The third is "quality blending laws," and in blending hydrocarbon liquids, the resulting properties or qualities are a usually a linear function of the components blended, except chemical interactions can cause inaccuracies in linear predictions thus requiring nonlinear "quality blending laws." The third is "purchase quantity discrete size restriction," and in many cases only specific quantities of a raw material can be purchased such as a railroad tank car of nitrogen tetra-oxide, a marine tanker of crude oil, and pipelines have specific minimum batch amounts including incremental delivery size orders. The fourth is "variable pool capacity and capacity conservation," and this includes cases where multipurpose equipment can perform one task at a time and limited storage tanks for multiple products.

Total Cost Assessment

The business focus of chemical companies has moved from a regional to a global basis, and this has redefined how these companies organize and view their activities. As described by H. J. Kohlbrand of Dow Chemical Company (Kohlbrand, 1998), the chemical industry has gone from end-of-pipe treatment to source reduction, recycling and reuse. Pollution prevention was an environmental issue and is now a critical business opportunity. Companies are undergoing difficult institutional transformations, and emphasis on pollution prevention has broadened to include tools such as Total (full) Cost Assessment (accounting) (TCA), Life Cycle Assessment (LCA), sustainable development and eco-efficiency (*eco*nomic and *eco*logical). There is no integrated set of tools, methodologies or programs to perform a consistent and accurate evaluation of new plants and existing processes. Some of these tools are available individually, e.g. TCA and LCA, and some are being developed, e.g. metrics for sustainability. An integrated analysis incorporating TCA, LCA and sustainability is required for proper identification of real, long- term benefits and costs that will result in the best list of prospects to compete for capital investment.

Chemical companies and petroleum refiners have applied total cost accounting and found that the cost of environmental compliance was three to five times higher than the original estimates (Constable, et. al., 2000). Total or full cost accounting identifies the real costs associated with a product or process. It organizes different levels of costs and includes direct, indirect, associated and societal. Direct and indirect costs include those associated with manufacturing. Associated costs include those associated with compliance, fines, penalties and future liabilities. Societal costs are difficult to evaluate since there is no standard, agreed-upon methods to estimate them, and they can include damage to the environment from emissions emitted within regulations, consumer response and employee relations, among others (Kohlbrand, 1998).

The Center for Waste Reduction Technology (CWRT) of the American Institute of Chemical Engineers (AIChE) completed a detailed report with an Excel spreadsheet on Total Cost Assessment Methodology (Constable, et al., 2000). This TCA report was the outgrowth of industry representatives working to develop the best methodology for use by the chemical industry. The AIChE/CWRT TCA program uses five types of costs. Type 1 costs are direct costs for the manufacturing site. Type 2 costs are potentially hidden corporate and manufacturing site overhead costs. Type 3 costs are future and contingent liability costs. Type 4 costs are internal intangible costs, and Type 5 costs are external costs that the company does not pay directly including those born by society and from deterioration of the environment by pollution within compliance regulations. This report states that environmental costs made up at least 22% of the non-feedstock operating costs of the Amoco's Yorktown oil refinery. Also, for one DuPont pesticide, environmental costs were 19% of the total manufacturing costs; and for one Novartis additive these costs were a minimum of 19% of manufacturing costs, excluding raw materials. In addition, this TCA methodology was said to have the capability to evaluate the full life cycle and consider environmental and health implications from raw material extraction to end-of-life of the process or product.

Sustainable development is the concept that development should meet the needs of the present without sacrificing the ability of the future to meet its needs. There have been many publications on sustainable development and environmental economics that have been described by Daly (Daly, 1996), and in 1995 the President's Council on Sustainable Development issued a report giving fifteen principles. These included calls: to preserve the integrity of natural systems, to have economic growth, environmental protection and social equity, to be interdependent; to have a stable population consistent with the carrying capacity of the earth, and to have all segments of society equally share environmental costs. How these principles will be considered and ways to proceed involve complex political, trade, health, and scientific and technical issues. Approaches have been and are being proposed by economists, government officials and business leaders. First, measures or metrics of sustainable development must be defined, tested and applied before sound policy decisions can be proposed and evaluated. An effort is underway to develop these metrics by an industry group through the Center for Waste Reduction of the American Institute of Chemical Engineers, and they have issued two interim reports (Adler, 1999) and held a workshop (Beaver and Beloff, 2000). Also, external or sustainable costs are the very difficult to quantify, and the TCA report gives some estimates for these costs from a study of environmental cost from pollutant discharge to air from electricity generation, e.g. $0.22-2.38 per ton for CO, 0-$3.25 per ton for CO_2.

Chemical Complex (Multi-Plant) Analysis System

The Chemical Complex Analysis System is a new methodology that has been developed to determine the best configuration of plants in a chemical complex based on economic, energy, environmental and sustainable costs. It is an integrated computer system that is used by plant and design engineers to convert their company's goals and capital into viable projects that are profitable and meet economic, energy and sustainability requirements. In addition, they can perform evaluations for impacts associated with green house gases and finite resources. These evaluations also can demonstrate that plants are delivering energy efficient, societal and business benefits that will help ameliorate command and control regulations. The program and users manual can be downloaded from the LSU Minerals Processing Research Institute's web site, http://www.mpri.lsu.edu/chemcomplex.html.

The program:
- Incorporates economic, energy, environmental and sustainable costs
- Solves for the optimum configuration of plants
- Incorporates EPA Pollution Index methodology
- Applied successfully to agricultural chemical complex
- Developed by university-industry team

Description: The Chemical Complex Analysis System structure is shown in Figure 31. It incorporates a flowsheeting component as shown in Figure 31 where simulations of the plants in the complex are entered. Each simulation includes process or block flow diagrams with material and energy balances, rates equations, equilibrium relations and thermodynamic and transports properties for the process units and heat exchanger networks. These equations are entered through windows and stored in the database to be shared with the other components of the system.

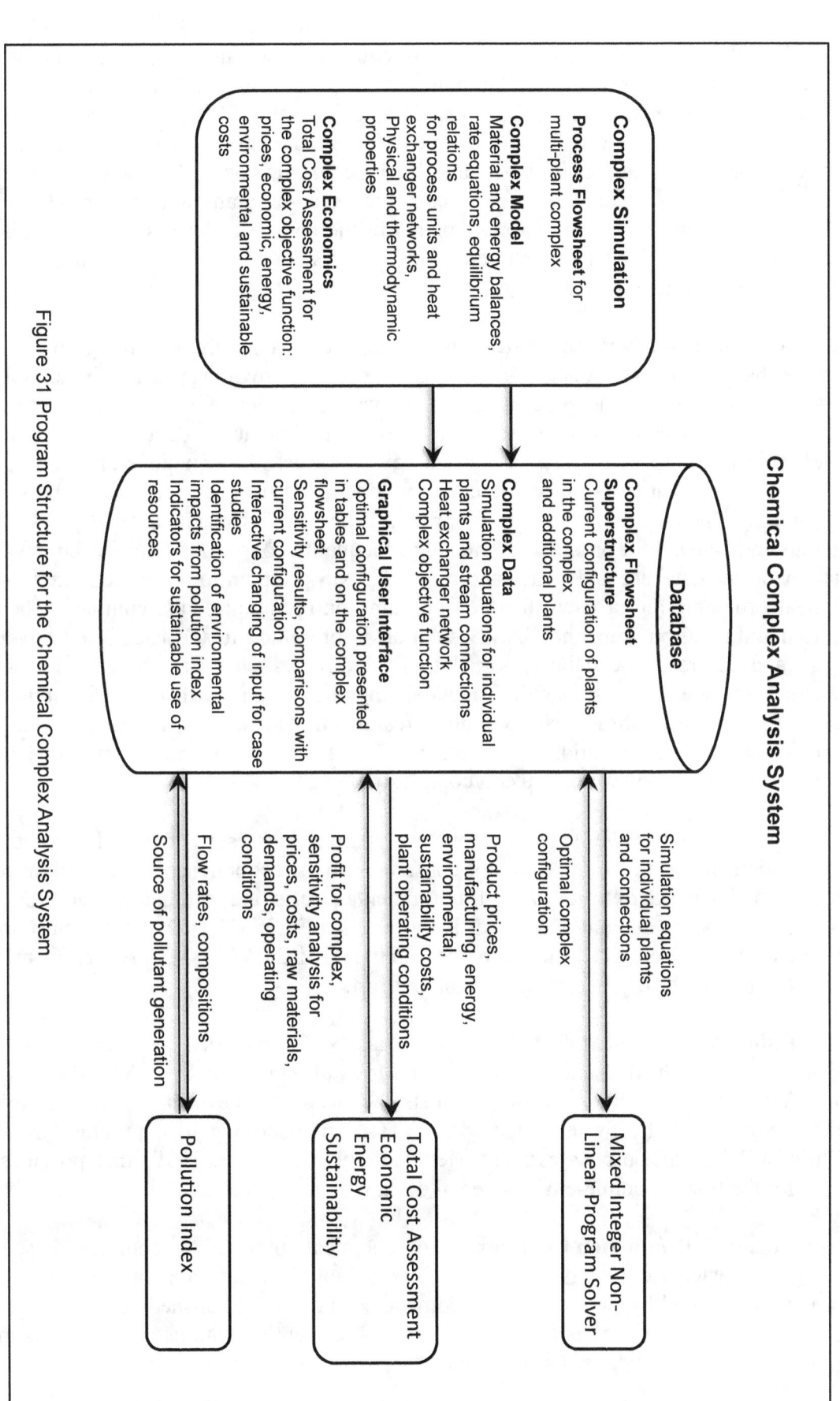

Figure 31 Program Structure for the Chemical Complex Analysis System

The objective function is entered as an equation associated with each process with related information for prices and economic, energy, environmental and sustainable costs that are used in the evaluation of the Total Cost Assessment (TCA) for the complex. The TCA includes the total profit for the complex that is a function of the economic, energy, environmental and sustainable costs and income from sales of products. Then the information is provided to the mixed integer nonlinear programming solver (GAMS) to determine the optimum configuration of plants in the complex. Also, sources of pollutant generation are located by the pollution index component of the system using the EPA pollution index methodology (Cabezas, et. al., 1997).

All interactions with the system shown in Figure 31 are through the graphical user interface that is written in Visual Basic. As the process flow diagram for the complex is prepared, equations for the process units and variables for the streams connecting the process units are entered and stored in the database using interactive data forms as shown on the left side in Figure 31. Material and energy balances, rate equations and equilibrium relations for the plants are entered as equality constraints using the format of the GAMS programming language that is similar to Fortran. Process unit capacities, availability of raw materials and demand for product are entered as inequality constraints. The System takes the equations in the database and writes and runs a GAMS program to solve the mixed integer nonlinear programming problem for the optimum configuration of the complex. Then the important information from the GAMS solution is presented to the user in a convenient format, and the results can be exported to Excel, if desired. Features for developing flowsheets include adding, changing and deleting the equations that describe units and streams and their properties. Usual Windows features include cut, copy, paste, delete, print, zoom, reload, update and grid, among others. A detailed description is provided in the user's manual (http://www.mpri.lsu.edu/complex.pdf).

The system has the TCA component prepare the assessment model for use with determination of the optimum complex configuration. Economic costs are estimated by standard methods (Garrett, 1989). Environmental costs are estimated from the data provided by Amoco, DuPont and Novartis in the AIChE/CWRT TCA report. Sustainable costs are estimated from the air pollution data in the AIChE/CWRT TCA report (Constable, 1999). Improving the estimates is an on-going effort.

Industry Collaboration: The system has been developed in collaboration with engineering groups at Monsanto Enviro Chem, Motiva Enterprises, IMC Agrico and Kaiser Aluminum and Chemicals to ensure it meets the needs of the chemical and petroleum refining industries. The System incorporates TCA methodology in a program from the AIChE/CWRT Total Cost Assessment Methodology (Constable, 1999) that provides the criteria for the best economic-environmental design.

Chemical Production Complex: A chemical production complex based on plants in the Baton Rouge - New Orleans, Mississippi river corridor was developed with information provided by the cooperating companies and other published sources, as shown in Figure 32. This complex is representative of the current operations and practices in the chemical industry and was used as the base case and starting point to develop a superstructure by adding plants. These additional plants gave alternate ways to produce intermediates that reduced and consumed wastes and greenhouse gases and conserved energy. They could provide combinations leading to a complex with lower environmental impacts and greater sustainability. The base case and additional plants form a superstructure that was evaluated using the economic, environmental and sustainable criteria to determine the optimum configuration of plants as described below.

As shown in Figure 32 the chemical production complex in the lower Mississippi River corridor has 13 production units plus associated utilities for power, steam and cooling water and facilities for waste treatment. A production unit contains more than one plant; and, for example, the sulfuric acid production unit contains five plants owned by two companies and the phosphoric acid production unit contains four plants owned by three companies. Here, ammonia plants produce 0.75 million metric tons/year of carbon dioxide, and methanol, urea, and acetic acid plants consume 0.14 million metric tons/year of carbon dioxide. This leaves a surplus of 0.61 million metric tons/year of high quality carbon dioxide, as shown in Figure 32. This high purity carbon dioxide is being vented to the atmosphere now.

The raw materials used in the base case of the chemical production complex include air, water, natural gas, sulfur, phosphate rock, ethylene and benzene as shown on Figure 32. The products are mono- and di-ammonium phosphate (MAP and DAP), granular triple super phosphate (GTSP), urea, ammonium nitrate, and urea ammonium nitrate solution (UAN), phosphoric acid, ammonia, methanol, acetic acid, ethylbenzene and styrene. The flow rates shown on the diagram are in million metric tons per year. Intermediates are sulfuric acid, phosphoric acid, ammonia, nitric acid, urea, carbon dioxide and ethylbenzene. The intermediates are used to produce MAP and DAP, GTSP, urea, ammonium nitrate, acetic acid, UAN, and styrene. Ammonia is used in direct application to crops and other uses. MAP, DAP, UAN and GTSP are used in direct application to crops. Phosphoric acid can be used in other industrial applications. Methanol is used to produce formaldehyde, methyl esters, amines and solvents, among others, and is included for its use of ammonia plant byproduct - carbon dioxide. Acetic acid, ethylbenzene and styrene are used as feedstock in other chemical processes. Emissions from the chemical production complex include sulfur dioxide, nitrogen oxides, ammonia, methanol, silicon tetrafluoride, hydrogen fluoride and gypsum.

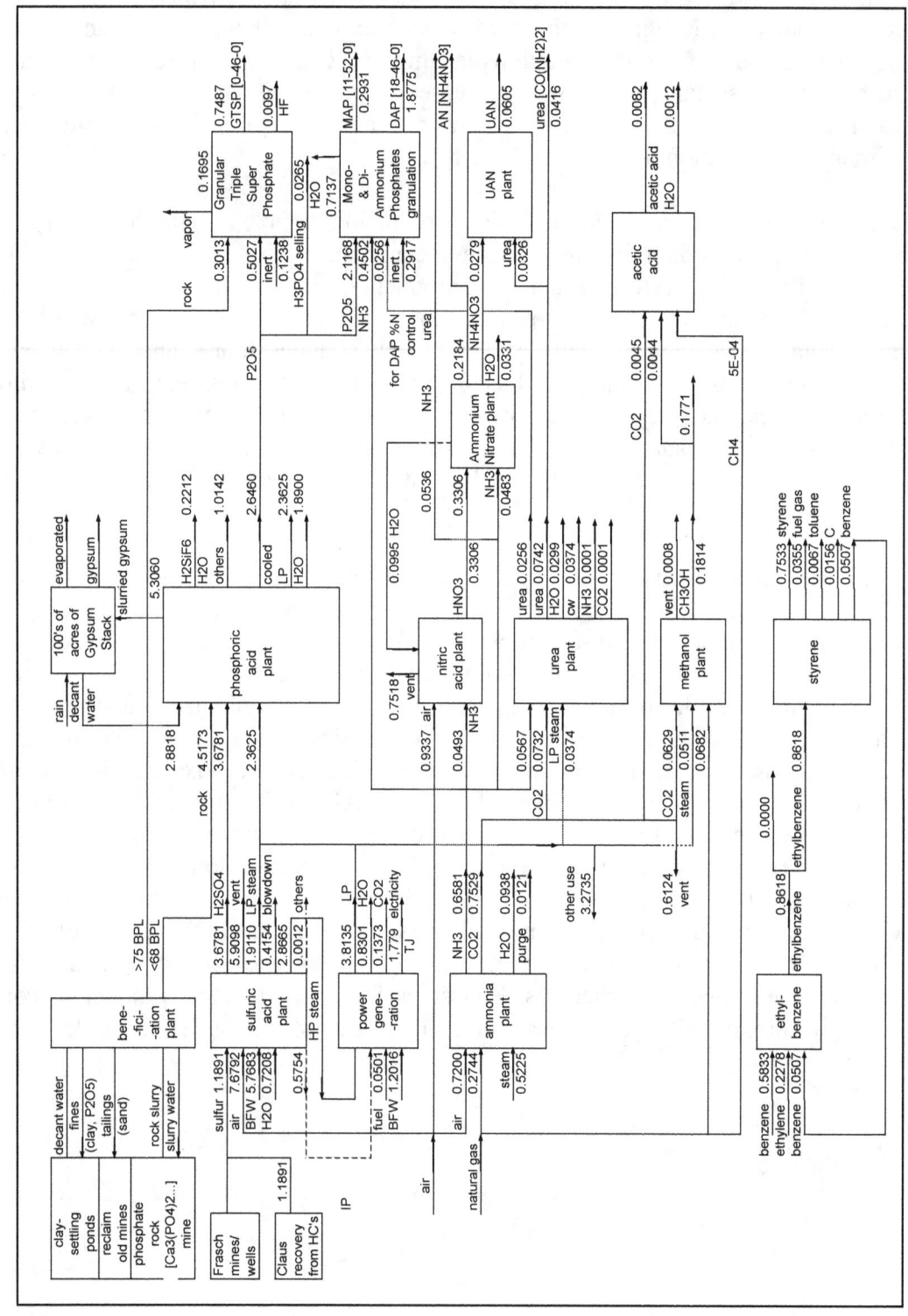

Figure 32 Chemical Production Complex in the Lower Mississippi River Corridor, Base Case Flow Rates in Million Metric Tons per Year

For this base case there was 270 equality constraint equations with 326 variables describing the material and energy balances and chemical conversions. Also, there were 28 inequality constraint equations describing the demand for product, availability of raw materials and range on the capacities of the individual plants in the complex and 56 degrees of freedom.

The chemical production complex shown in Figure 32 was expanded into a superstructure as shown in Table 9 and Figure 33 by adding 18 new chemical processes. Several approaches were incorporated in this expanded complex with alternate ways to produce intermediates that reduce wastes and energy and consume greenhouse gases. These plants consumed the excess carbon dioxide being vented currently and gave alternative ways to produce phosphoric acid, and recover sulfur and sulfur dioxide from gypsum waste.

Fourteen potentially new processes for consuming CO_2 were selected and integrated into the superstructure based on the evaluations of HYSYS simulations (Indala, 2004). These processes include four processes for methanol production, two processes for propylene, and one process each for ethanol, DME, formic acid, acetic acid, styrene, methylamines, graphite and H_2, and synthesis gas.

Four other new processes that do not use CO_2 as a raw material were incorporated in the superstructure. Two additional plants were added to produce phosphoric acid. One is the electric furnace process, which has high-energy cost but produce phosphoric acid. In the other process (Haifa process), calcium phosphate ore reacts with hydrochloric acid to produce phosphoric acid. Also, there are two plants that use gypsum wastes to recover sulfur and sulfur dioxide. One reduces gypsum waste to sulfur dioxide that is recycled to the sulfuric acid plant. The other reduces gypsum waste to sulfur and sulfur dioxide that are recycled to the sulfuric acid plant. Thus, a total of eighteen new processes were included in the superstructure.

The superstructure had 735 continuous variables, 23 integer variables, and 601 equality constraints that describe material and energy balances for the plants. Also, there were 77 inequality constraints that describe availability of raw materials, demand for products, capacities of the plants, and logical relations in the chemical complex. The degrees of freedom were 134, and the optimal solution obtained with the Chemical Complex Analysis System is discussed below.

Table 9 Processes in Chemical Production Complex Base Case and Superstructure, Xu, 2000

Plants in the Base Case	Plants Added to Form the Superstructure
Ammonia	Methanol - Bonivardi, et al., 1998
Nitric acid	Methanol – Jun, et al., 1998
Ammonium nitrate	Methanol – Ushikoshi, et al., 1998
Urea	Methanol – Nerlov and Chorkendorff, 1999
UAN	Ethanol
Methanol	Dimethyl ether
Granular triple super phosphate (GTSP)	Formic acid
MAP and DAP	Acetic acid - new method
Contact process for sulfuric acid	Styrene - new method
Wet process for phosphoric acid	Methylamines
Acetic acid – conventional method	Graphite
Ethyl benzene	Hydrogen/Synthesis gas
Styrene	Propylene from CO_2
Power generation	Propylene from propane dehydrogenation
	Electric furnace process for phosphoric acid
	Haifa process for phosphoric acid
	SO_2 recovery from gypsum waste
	S and SO_2 recovery from gypsum waste

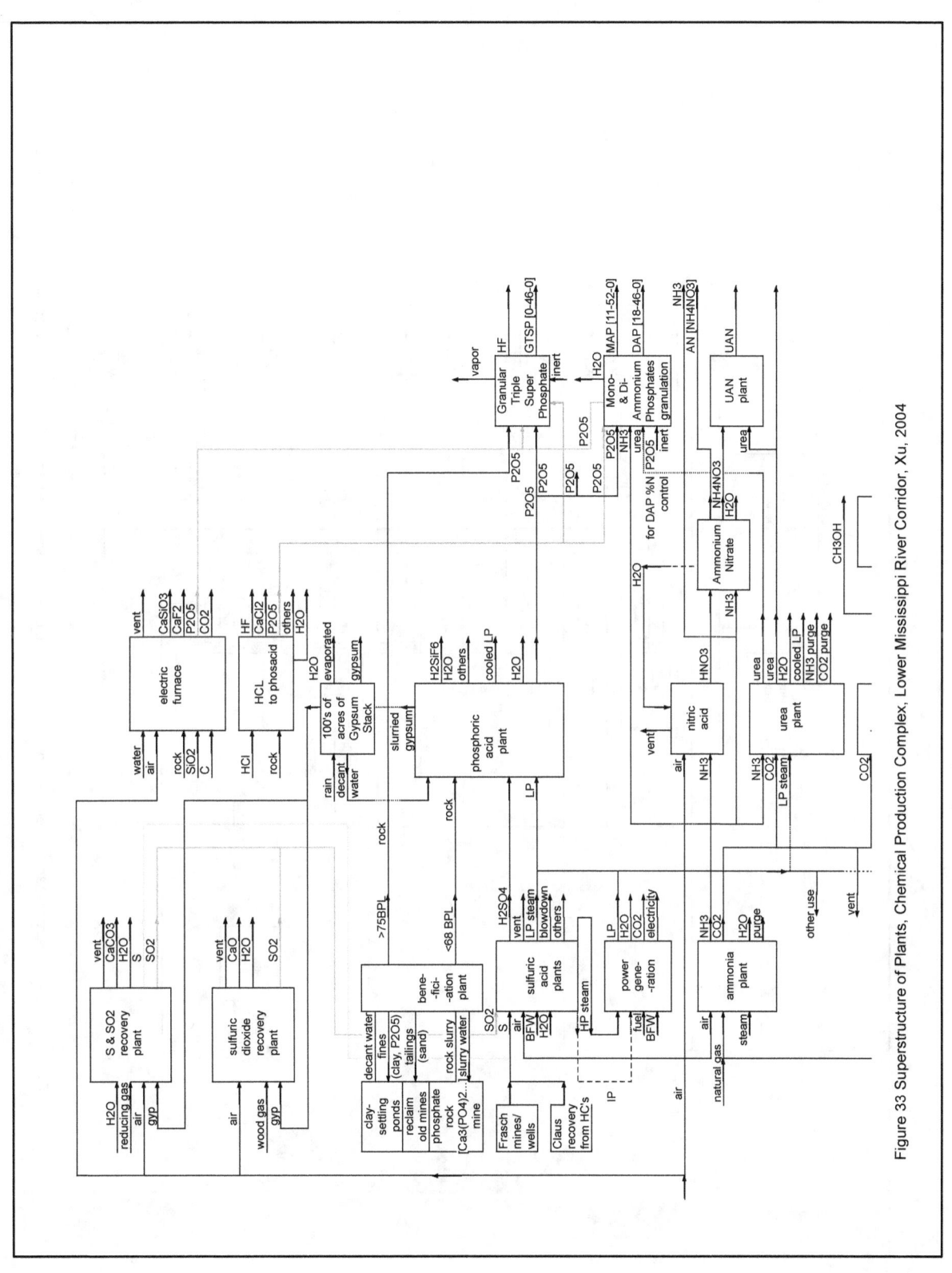

Figure 33 Superstructure of Plants, Chemical Production Complex, Lower Mississippi River Corridor, Xu, 2004

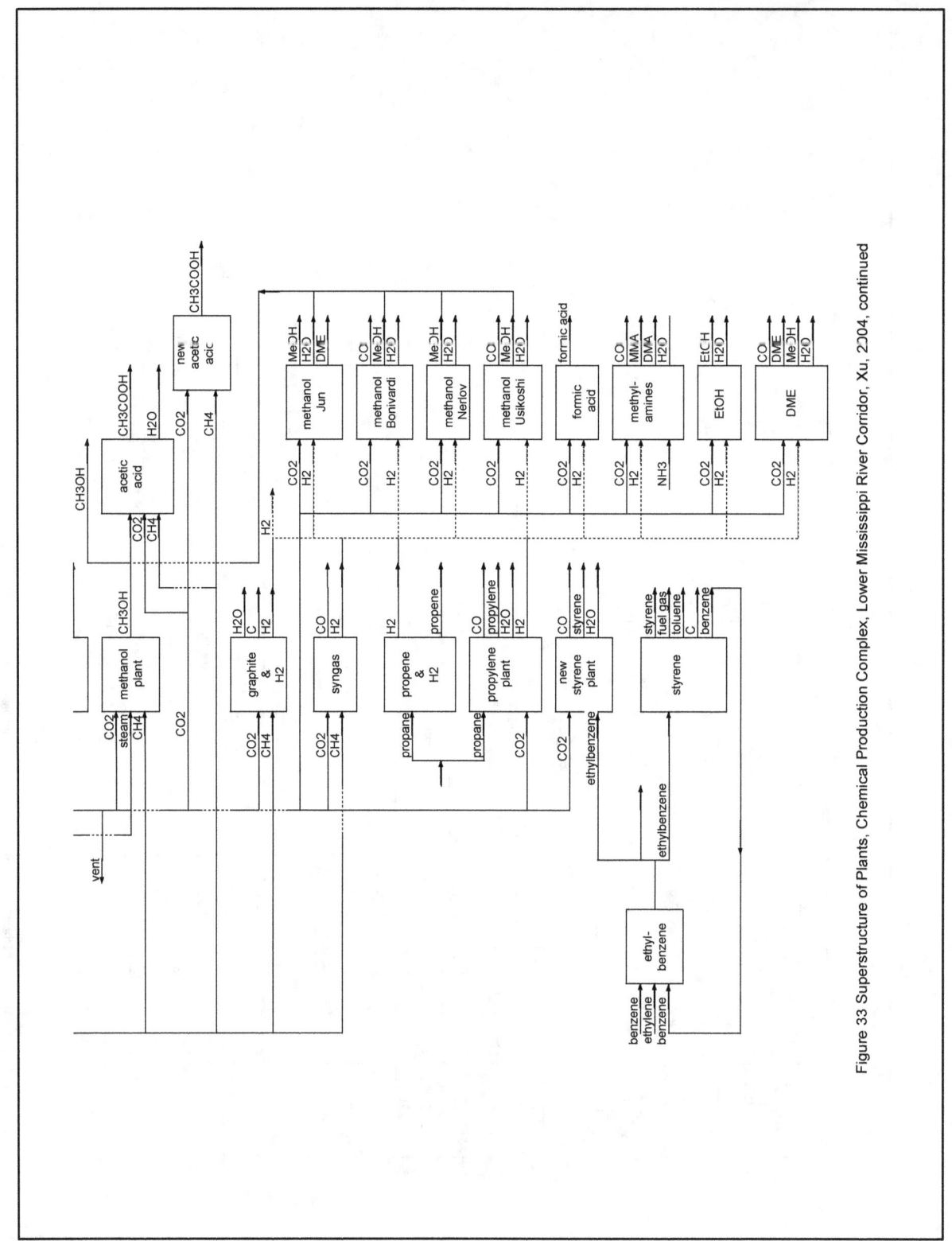

Figure 33 Superstructure of Plants, Chemical Production Complex, Lower Mississippi River Corridor, Xu, 2004, continued

102

Triple Bottom Line: To determine the optimum configuration of plants in the chemical complex, a value added economic model was used for the base case, and it is the difference between sales and the cost of raw materials and assumes other manufacturing costs are constant. The value-added economic model is given by the profit in Equation 49, and it was expanded to account for environmental and sustainable costs. Environmental costs are costs required to comply with federal and state environmental regulations including permits, monitoring emissions, fines, etc., as described in the AIChE/TCA report (Constable, et al., 2000). Sustainable costs are costs to society from damage to the environment by emissions discharged within permitted regulations. This extended value-added economic model is referred to as the "triple bottom line" and is the difference between sales and sustainable credits and economic costs (raw materials and utilities), environmental costs and sustainable costs as given by Equation 50. The sales prices for products and the costs of raw materials are given in Table 10 along with sustainable costs and credits. Also, the standard deviations of the prices and costs are given in Table 10 that were used in the sensitivity analysis.

$$\text{Profit} = \Sigma \text{ Product Sales} - \Sigma \text{ Raw Material Costs} - \Sigma \text{ Energy Costs} \qquad (49)$$

$$\text{Triple Bottom Line} = \Sigma \text{ Product Sales} - \Sigma \text{ Raw Material Costs} - \Sigma \text{ Energy Costs}$$

$$- \Sigma \text{ Environmental Costs} + \Sigma \text{ Sustainable (Credits} - \text{Costs)}$$

$$\text{Triple Bottom Line} = \text{Profit} - \Sigma \text{ Environmental Costs} + \Sigma \text{ Sustainable (Credits} - \text{Costs)} \quad (50)$$

Environmental costs were estimated to be 67% of the raw material costs based on the data provided by Amoco, DuPont and Novartis in the AIChE/TCA report (Constable, et al., 2000). This report lists environmental costs as approximately 20% of the total manufacturing costs and raw material costs as approximately 30% of total manufacturing costs.

Sustainable costs were estimated from results given for power generation in the AIChE/TCA report where CO_2 emissions had a sustainable cost of \$3.25 per metric ton of CO_2. As shown in Table 10, a cost of \$3.25 was charged as a cost to plants that emitted CO_2, and a credit of twice this cost (\$6.50) was given to plants that utilized CO_2. This credit was included for steam produced from waste heat by the sulfuric acid plant displacing steam produced from a package boiler firing hydrocarbons and emitting carbon dioxide. In this report SO_2 and NO_X emissions had sustainable costs of \$192 per metric ton of SO_2 and \$1,030 per metric ton of NO_X. In addition, for gypsum production and use, an arbitrary but conservative sustainable cost of \$2.5 per metric ton for gypsum production was used, and a credit of \$5.0 per metric ton for gypsum consumption was used.

Table 10 Raw Material Costs, Product Prices and Sustainable Costs and Credits, Source: Constable, et al. (2000), Chemical Market Reporter, Camford Chemical Prices, Internet and C&EN (2003)

Raw Materials	Cost ($/mt)	Standard deviation ($/mt)	Sustainable Cost and Credits	Cost /Credit ($/mt)	Products	Price ($/mt)	Standard deviation ($/mt)
Natural gas	235	69.4	Credit for CO_2 consumption	6.50	Ammonia	224	17.7
Phosphate rock	27	-	Debit for CO_2 production	3.25	Methanol	271	43.2
Wet process	34	-	Debit for NO_x production	1,030	Acetic acid	1,030	36.6
Electric furnace	34	-	Debit for SO_2 production	192	GTSP	132	-
Haifa process	34	-	Credit for gypsum consumption	5.0	MAP	166	4.20
GTSP process	32	-	Debit for gypsum production	2.5	DAP	179	7.89
HCl	95	11.1			NH_4NO_3	146	6.66
Sulfur					Urea	179	17.4
Frasch	53	9.50			UAN	120	-
Claus	21	3.55			Phosphoric acid	496	-
Coke electric furnace	124	-			Hydrogen	1,030	252
Propane	180	-			Ethylbenzene	556	75.9
Benzene	303	60.3			Styrene	824	94.7
Ethylene	565	95.4			Propylene	519	66.0
Reducing gas	75	-			Formic acid	690	-
Wood gas	88	-			MMA	1,610	-
					DMA	1,610	-
					DME	946	-
					Ethanol	933	23.1
					Toluene	384	47.7
					Graphite	904	82.0
					Fuel gas	784	-
					CO	45	13.3

Optimal Configuration of Plants: The Chemical Complex Analysis System was used to obtain the optimum configuration of plants from the superstructure with the complete solution given by Xu, 2004. The optimum configuration of plants was obtained from the superstructure by maximizing the triple bottom line, Equation 50, subject to the equality and inequality constraints from the material and energy balances and related equations.

The optimal structure from the superstructure is shown in Figure 34, and a convenient way to show the new plants in the optimal structure is given in Table 11. Seven new processes in the optimal structure were selected from eighteen new processes in the superstructure. These included acetic acid, graphite, formic acid, methylamines, propylene (2) and synthesis gas production. The new acetic acid process replaced the commercial acetic acid plant in the chemical complex. The processes for dimethyl ether, styrene, and methanol were not selected in the optimal structure. It was more profitable to have the corresponding commercial processes present. The commercial process for methanol does not use expensive hydrogen as a raw material, but the new methanol processes does.

Table 11 Plants in the Optimal Structure from the Superstructure

Existing Plants in the Optimal Structure	New Plants in the Optimal Structure
Ammonia	Formic acid
Nitric acid	Acetic acid – new process
Ammonium nitrate	Methylamines (MMA and DMA)
Urea	Graphite
UAN	Hydrogen/synthesis gas
Methanol	Propylene from CO_2
Granular triple super phosphate (GTSP)	Propylene from propane dehydrogenation
MAP and DAP	
Contact process for Sulfuric acid	New Plants Not in the Optimal Structure
Wet process for phosphoric acid	Methanol – Bonivardi, et al., 1998
Ethylbenzene	Methanol – Jun, et al., 1998
Styrene	Methanol – Ushikoshi, et al., 1998
Power generation	Methanol – Nerlov and Chorkendorff, 1999
	Ethanol
Existing Plants Not in the Optimal Structure	Dimethyl ether
	Styrene - new method
Acetic acid	Electric furnace process for phosphoric acid
	Haifa process for phosphoric acid
	SO_2 recovery from gypsum waste
	S and SO_2 recovery from gypsum waste

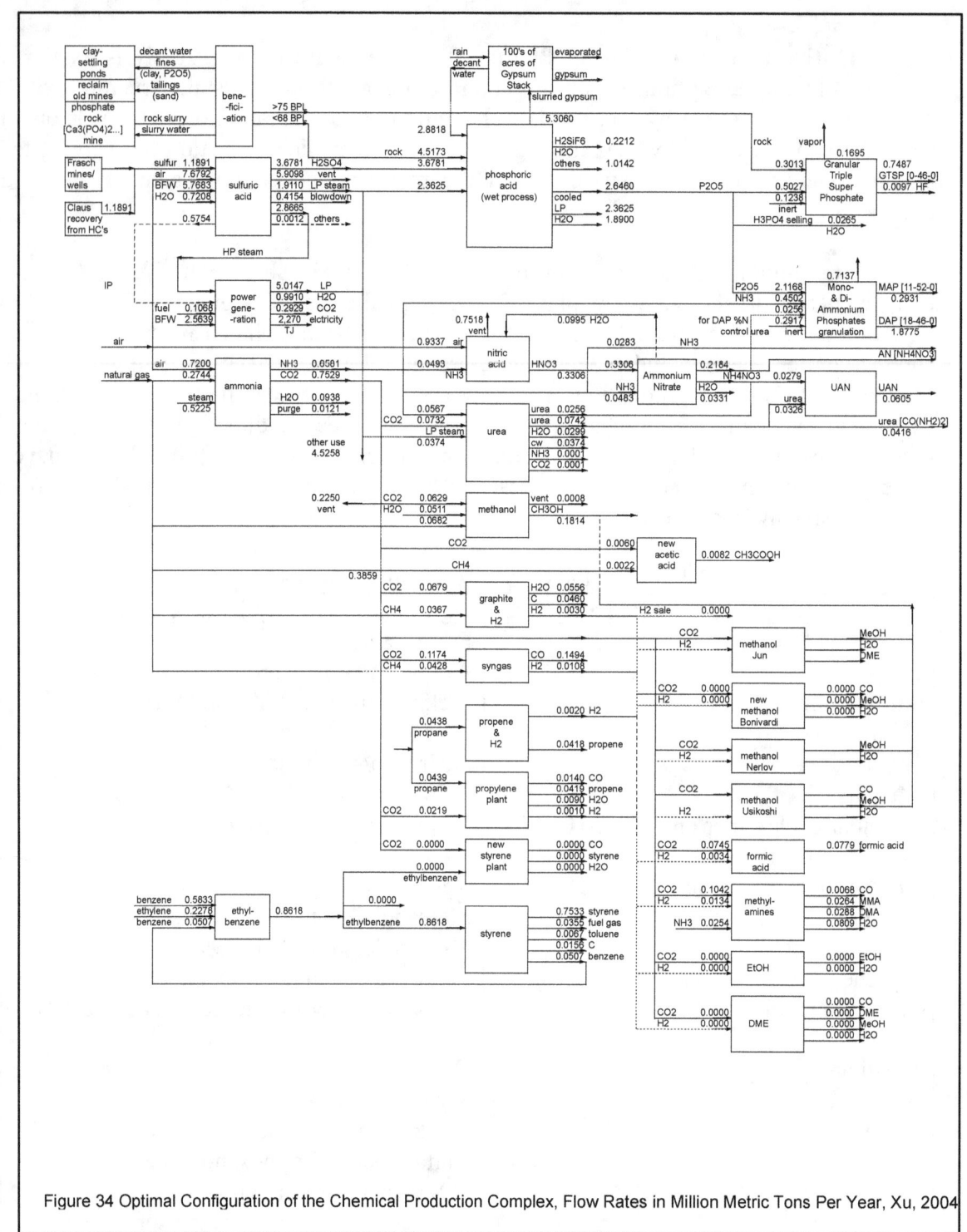

Figure 34 Optimal Configuration of the Chemical Production Complex, Flow Rates in Million Metric Tons Per Year, Xu, 2004

Comparison of the sales and costs associated with the triple bottom line, Equation 50, are shown in Table 12 for the base case and the optimal structure. The triple bottom line increased from $343 to $506 million per year or about 48% from the base case to the optimal structure. Sales increased from additional products from carbon dioxide, and there were corresponding increases in the other costs associated with producing these products by the companies. Cost to society has improved since sustainable costs decreased from $18 to $15 million per year from the credits given for using carbon dioxide and increased energy efficiency.

Table 12 Sales and Costs Associated with the Triple Bottom Line for the Base Case and Optimal Structure

	Base Case million dollars/year	Optimal Structure million dollars/year
Income from Sales	1,277	1,508
Economic Costs (Raw Materials and Utilities)	554	602
Raw Material Costs	542	577
Utility Costs	12	25
Environmental Cost (67% of Raw Material Cost)	362	385
Sustainable Credits (+)/Costs (-)	-18	-15
Triple Bottom Line	343	506

The increased use of carbon dioxide is shown in Table 13 for the optimal structure. However, it was not optimal to consume all of the carbon dioxide available, and 0.22 million metric tons per year is vented to the atmosphere, down by 0.39 million metric tons per year or 64%.

Table 13 Carbon Dioxide Consumption in Bases Case and Optimal Structure

	Base Case million metric tons/year	Optimal Structure million metric tons/year
CO_2 produced by NH_3 plant	0.75	0.75
CO_2 consumed by methanol, urea and other plants	0.14	0.53
CO_2 vented to atmosphere	0.61	0.22

Six of the seven new processes present in the optimal structure use CO_2 as a raw material as shown in Tables 11 and 14. In Table 14, the optimal capacities are given for the plants in the optimum structure of the chemical production complex. Also shown in this table is the energy used or produced for each process and the total energy required for the complex. With the additional plants in the optimal structure, the energy required increased from 2,150 to 5,791 TJ/year. This is reflected in the increased utility cost shown in Table 12

107

Table 14 Comparisons of Capacities for the Base Case and Optimal Structure

Plant name	Capacity (upper-lower bounds) (mt/year)	Base case Capacity (mt/year)	Energy Requirement (TJ/year)	Optimal Capacity (mt/year)	Energy Requirement (TJ/year)
Ammonia	329,000-658,000	658,000	3,820	658,000	3,820
Nitric acid	89,000-178,000	178,000	-775	178,000	-775
Ammonium nitrate	113,000-227,000	227,000	229	227,000	229
Urea	49,900-99,800	99,800	128	99,800	128
Methanol	91,000-181,000	181,000	2,165	181,000	2,165
UAN	30,000-60,000	60,000	0	60,000	0
MAP	146,000-293,000	293,000		293,000	
DAP	939,000-1,880,000	1,880,000	1,901	1,880,000	1,901
GTSP	374,000-749,000	749,000	1,312	749,000	1,312
Sulfuric acid	1,810,000-3,620,000	3,620,000	-14,642	3,620,000	-14,642
Wet process phosphoric acid	635,000-1,270,000	1,270,000	5,181	1,270,000	5,181
Ethylbenzene	431,000-862,000	862,000	-755	862,000	-755
Styrene	386,000-771,000	753,000	3,318	753,000	3,318
Acetic acid	4,080-8,160	8,160	268	0	0
Electric furnace phosphoric acid	635,000-1,270,000	na	na	0	0
Haifa phosphoric acid	635,000-1,270,000	na	na	0	0
New Acetic acid	4,090-8,180	na	na	8,180	8
SO_2 recovery from gypsum	987,000-1,970,000	na	na	0	0
S and SO_2 recovery from gypsum	494,000-988,000	na	na	0	0
Graphite and H_2 from CO_2 and CH_4	230,000-460,000	na	na	46,000	1,046
Syngas	6,700-13,400	na	na	10,800	691
Propene and H_2	20,900-41,800	na	na	41,800	658
Propene using CO_2	21,000-41,900	na	na	41,900	413
New Styrene	181,000-362,000	na	na	0	0
New methanol – Bonivardi	240,000-480,000	na	na	0	0
New methanol – Jun	240,000-480,000	na	na	0	0

Table 14 continued

Plant name	Capacity (upper-lower bounds) (mt/year)	Base case Capacity (mt/year)	Energy Requirement (TJ/year)	Optimal Capacity (mt/year)	Energy Requirement (TJ/year)
New methanol – Nerlov	240,000-480,000	na	na	0	0
New methanol – Ushikoshi	240,000-480,000	na	na	0	0
Formic acid	39,000-78,000	na	na	78,000	14
Methylaimines	13,200-26,400	na	na	26,400	1,079
Ethanol	52,000-104,000	na	na	0	0
Dimethyl ether	22,900-45,800	na	na	0	0
Ammonia sale		53,600		28,300	
Ammonium Nitrate sale		218,000		218,000	
Urea sale		41,600		41,600	
Wet process phosphoric acid sale		12,700		12,700	
Ethylbenzene sale		0		0	
CO_2 vented		612,000		225,000	
Total energy requirement			2,150		5,791

going from \$12 to \$25 million per year. This additional energy is supplied from firing boilers with natural gas that has a sustainable cost of \$3.25 per metric ton. As shown in Table 14 the sulfuric acid plant is an important source of energy as steam, and operating this plant for steam production is as important as production of sulfuric acid (Xu, 2004).

Multiobjective or Multicriteria Optimization of a Chemical Production Complex: The objective is to find optimal solutions that maximize companies' profits and minimize costs to society. Companies' profits are sales minus economic and environmental costs. Economic costs include raw material, utilities, labor, and other manufacturing costs. Environmental costs include permits, monitoring of emissions, fines, etc. The costs to society are measured by sustainable costs, and these costs are from damage to the environment by emissions discharged within permitted regulations. Sustainable credits are awarded for reductions in emissions as shown in Table 10, and they are similar to emissions trading credits.

The multicriteria optimization problem can be stated as in terms of profit, P, and sustainable credits/costs, S, for theses two objectives in Equation 51.

$$\text{Max:} \quad P = \Sigma \text{ Product Sales} - \Sigma \text{ Economic Costs} - \Sigma \text{ Environmental Costs} \qquad (51)$$
$$S = \Sigma \text{ Sustainable (Credits} - \text{Costs)}$$

Subject to: Multi-plant material and energy balances
Product demand, raw material availability, plant capacities

Multicriteria optimization obtains solutions that are called efficient or Pareto optimal solutions. These are optimal points where attempting to improving the value of one objective would cause another objective to decrease. To locate Pareto optimal solutions, multicriteria optimization problems are converted to one with a single criterion by a parametric approach method, which is by applying weights to each objective and optimizing the sum of the weighted objectives. The multicriteria mixed integer optimization problem becomes:

$$\text{Max:} \qquad w_1 P + w_2 S \qquad\qquad\qquad\qquad (52)$$

Subject to: Multi-plant material and energy balances
Product demand, raw material availability, plant capacities

The Chemical Complex Analysis System was used to determine the Pareto optimal solutions for the weights using $w_1 + w_2 = 1$ given in Equation 52, and these results are shown in Figure 35. Company profits are an order of magnitude larger than sustainable credits/costs. Sustainable credits/costs decline and company profits increase as the weight, w_1, on company profits increase. For example, when $w_1 = 1$, this is the optimal solution shown in Table 15 for P = \$520.6 and S = \$-14.76 million per year. The optimal solution with $w_1 = 0$ gave P = \$94.37 and S = \$23.24 million per year.

The points shown in Figure 35 are the Pareto optimal solutions for w_1 from 0 to 1.0 for increments of 0.001 from Xu, 2004. The values for w_1 equal to 0 and 1.0 and some intermediate ones are shown in Table 15. It shows that the sustainable costs become credits of \$0.68 million per year for a profit of \$389.8 million per year.

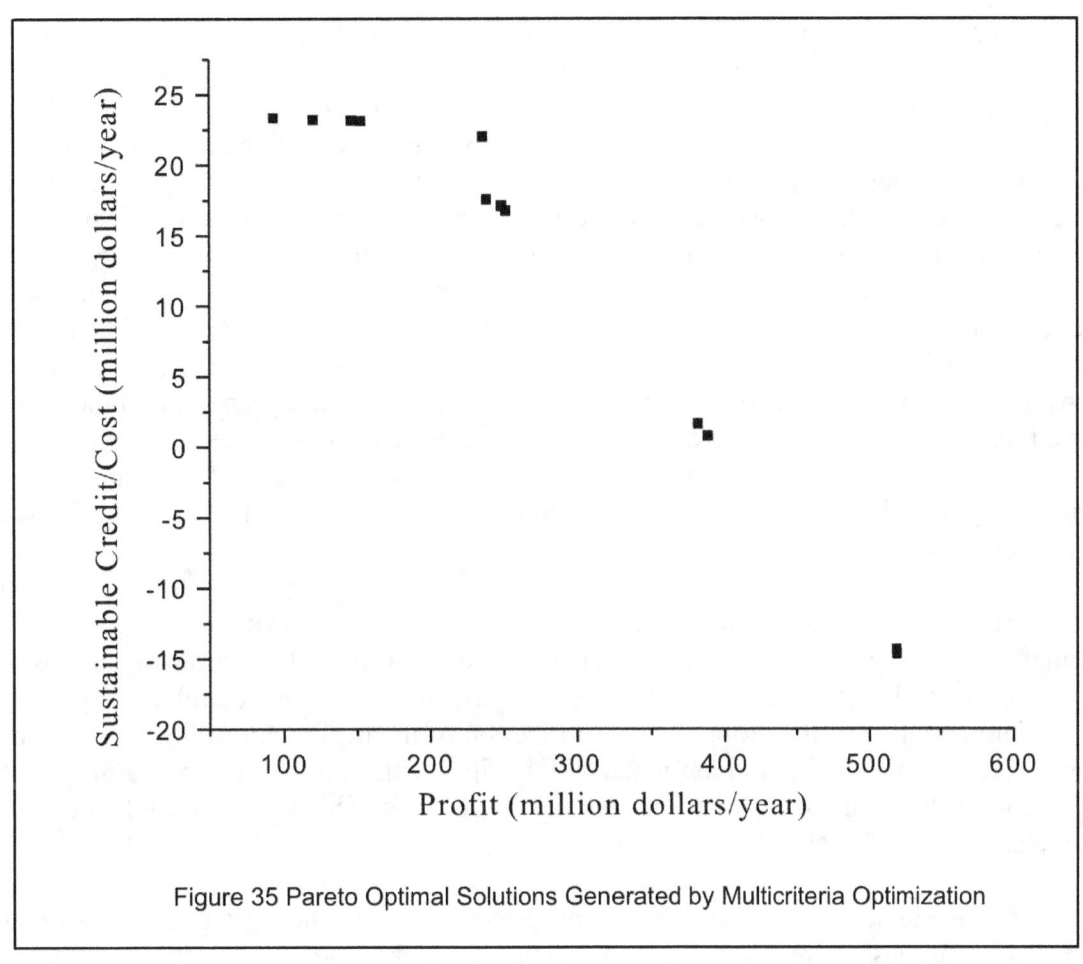

Figure 35 Pareto Optimal Solutions Generated by Multicriteria Optimization

Table 15 Several Values of the Pareto Optimal Solutions shown in Figure 35

Profit (million dollars/year)	Sustainable Credits/Costs (million dollars/year)	Weight (w_1)
520.6	-14.76	1
520.1	-14.45	0.417
389.8	0.68	0.172
383.1	1.53	0.115
252.9	16.68	0.104
249.8	17.02	0.103
239.7	17.48	0.039
237.1	21.93	0.037
153.7	23.04	0.013
147.6	23.07	0.004
121.5	23.12	0.002
94.37	23.24	0

The chemical production complex configurations of the Pareto optimal solutions for w_1 from 0 to 1.0 for increment of 0.001 are shown in Table 16. If a process is selected, the binary variable associated with the process is 1, otherwise 0. For each processes in Table 16, the sums of the binary variable values for the corresponding w_1 range are shown, along with the total summation of the times the process was selected. New acetic acid process always replaced the conventional one. The conventional methanol always operated instead of the four potentially new methanol processes. Synthesis gas, formic acid, propylene from CO_2, propylene from propane dehydrogenation, graphite, wet process phosphoric acid, and ethyl benzene process always operated. Ethanol, electric furnace phosphoric acid, and Haifa process phosphoric acid never operated. Only when w_1 was very small (0 - 0.150), SO_2 recovery from gypsum, S and SO_2 recovery from gypsum, new styrene, and dimethyl ether started operation. Methylamines and styrene processes always ran except that when w_1 was very small (0-0.150). Hence, the optimal structure is affected, but it did not change significantly (Table 16). It is another decision to determine the specific value of the weight that is acceptable to all concerned.

Monte Carlo Simulation using the Chemical Production Complex: Monte Carlo simulation was used to determine the sensitivity of the optimal solution to the costs and prices used in the triple bottom line. One of the results is the cumulative probability distribution, a curve of the probability as a function of the triple bottom line. A value of the cumulative probability for a given value of the triple bottom line is the probability that the triple bottom line will be equal to or less than that value. This curve is used to determine upside and downside risks.

For a Monte Carlo simulation, mean prices and costs along with estimates of their standard deviations are required. The costs and prices in Table 10 were used, and standard deviations estimated from cost and price fluctuations from the sources were given in Table 10 over a three- to five-year period. Sustainable costs and credits were constant, and sensitivity to these values is to be determined in a subsequent evaluation.

A total of 1,000 Monte Carlo simulations were selected based on two methods, uncertainty about the mean and confidence interval for fractals, for a 95% confidence interval (Xu, 2004).

Table 16 Optimal Structure Changes in Multicriteria Optimization (Number of Times out of 1,001 a Process is Selected)

Processes	w_1							
	0.000-0.149	0.150-0.299	0.300-0.449	0.450-0.599	0.600-0.749	0.750-0.899	0.900-1.000	Sum
Electric furnace phosphoric acid (Y_1)	0	0	0	0	0	0	0	0
Acetic acid (Y_{11})	0	0	0	0	0	0	0	0
New acetic acid (Y_{12})	150	150	150	150	150	150	101	1,001
SO_2 recovery from gypsum (Y_{13})	112	23	0	0	0	0	0	135
S and SO_2 recovery from gypsum (Y_{14})	38	0	0	0	0	0	0	38
Methanol (Y_{16})	150	150	150	150	150	150	101	1,001
Haifa process phosphoric acid (Y_2)	0	0	0	0	0	0	0	0
Propylene from CO_2 (Y_{23})	150	150	150	150	150	150	101	1,001
Propylene from propane dehydrogenation (Y_{24})	150	150	150	150	150	150	101	1,001
Synthesis gas (Y_{27})	150	150	150	150	150	150	101	1,001
Formic acid (Y_{29})	150	150	150	150	150	150	101	1,001
Wet process phosphoric acid (Y_3)	150	150	150	150	150	150	101	1,001
Methylamines (Y_{30})	149	150	150	150	150	150	101	1,000
Methanol (Jun, et al., 1998) (Y_{31})	0	0	0	0	0	0	0	0
Methanol (Bonivardi, et al., 1998) (Y_{32})	0	0	0	0	0	0	0	0
Methanol (Nerlov and Chorkendorff, 1999) (Y_{33})	0	0	0	0	0	0	0	0
Methanol (Ushikoshi, et al., 1998) (Y_{34})	0	0	0	0	0	0	0	0

Table 16 Continued

Processes	w_1							
	0.000-0.149	0.150-0.299	0.300-0.449	0.450-0.599	0.600-0.749	0.750-0.899	0.900-1.000	Sum
New styrene (Y_{35})	14	0	0	0	0	0	0	14
Ethanol (Y_{37})	0	0	0	0	0	0	0	0
Dimethyl ether (Y_{38})	40	0	0	0	0	0	0	40
Graphite (Y_{39})	150	150	150	150	150	150	101	1,001
Styrene (Y_{40})	136	150	150	150	150	150	101	987
Ethyl benzene (Y_{41})	150	150	150	150	150	150	101	1,001

The cumulative probability distribution shown in Figure 36 was obtained from the total of 1,000 Monte Carlo simulations. The mean for the triple bottom line was $513 million/year, and the standard deviation was $109 million/year. For the 1,000 samples, the maximum was $901 million/year, and the minimum was $232 million/year. A value of the cumulative probability for a given value of the triple bottom line is the probability that the triple bottom line will be equal to or less that value. For example, interpolated from Figure 36, there is 50% probability that the profit is equal to or less than $510 million per year.

The chemical production complex configurations of Monte Carlo simulation solutions for 1,000 samples are shown in Table 17. If a process is selected, the binary variable associated with the process is 1, otherwise 0. For each processes in Table 17, the sums of the binary variable values for the corresponding iteration range are shown, along with the total summation of the times the process was selected. New acetic acid process always replaced the conventional one. The conventional methanol almost always operated instead of the four potentially new methanol processes. The conventional styrene almost always operated instead of the potentially new styrene process. Synthesis gas, formic acid, propylene from CO_2, graphite, wet process phosphoric acid, and methylamines process always operated. Ethanol, electric furnace phosphoric acid, Haifa process phosphoric acid, SO_2 recovery from gypsum, S and SO_2 recovery from gypsum, dimethyl ether, and four new methanol processes never operated. New styrene process only started operation twice out of one thousand iterations. Ethyl benzene, and propylene from propane dehydrogenation almost always ran. Hence, the optimal structure is affected, but it did not change significantly (Table 17).

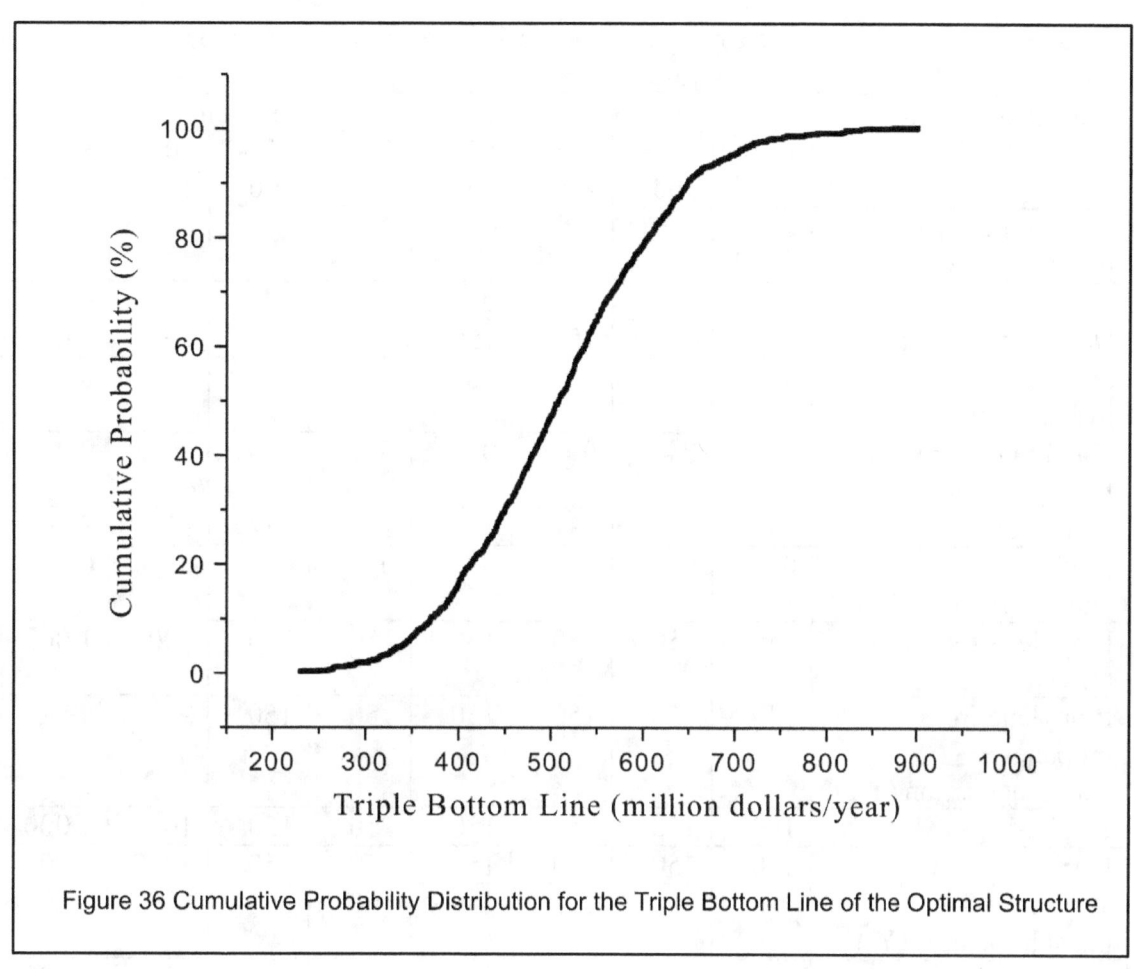

Figure 36 Cumulative Probability Distribution for the Triple Bottom Line of the Optimal Structure

Table 17 Optimal Structure Changes in Monte Carlo Simulation (Number of Times out of 1,000 a Process is Selected) from Xu, 2004

Processes	Monte Carlo Simulation (Iterations)							
	1-150	151-300	301-450	451-600	601-750	751-900	901-1000	Sum
Electric furnace phosphoric acid (Y_1)	0	0	0	0	0	0	0	0
Acetic acid (Y_{11})	0	0	0	0	0	0	0	0
New acetic acid (Y_{12})	150	150	150	150	150	150	100	1,000
SO_2 recovery from gypsum (Y_{13})	0	0	0	0	0	0	0	0
S and SO_2 recovery from gypsum (Y_{14})	0	0	0	0	0	0	0	0
Methanol (Y_{16})	138	133	131	129	125	132	85	873
Haifa process phosphoric acid (Y_2)	0	0	0	0	0	0	0	0
Propylene from CO_2 (Y_{23})	150	150	150	150	150	150	100	1,000
Propylene from propane dehydrogenation (Y_{24})	149	150	150	150	150	150	99	998
Synthesis gas (Y_{27})	150	150	150	150	150	150	100	1,000
Formic acid (Y_{29})	150	150	150	150	150	150	100	1,000
Wet process phosphoric acid (Y_3)	150	150	150	150	150	150	100	1,000
Methylamines (Y_{30})	150	150	150	150	150	150	100	1,000
Methanol (Jun, et al., 1998) (Y_{31})	0	0	0	0	0	0	0	0
Methanol (Bonivardi, et al., 1998) (Y_{32})	0	0	0	0	0	0	0	0
Methanol (Nerlov and Chorkendorff, 1999) (Y_{33})	0	0	0	0	0	0	0	0
Methanol (Ushikoshi, et al., 1998) (Y_{34})	0	0	0	0	0	0	0	0
New styrene (Y_{35})	0	1	1	0	0	0	0	2
Ethanol (Y_{37})	0	0	0	0	0	0	0	0
Dimethyl ether (Y_{38})	0	0	0	0	0	0	0	0
Graphite (Y_{39})	150	150	150	150	150	150	100	1,000
Styrene (Y_{40})	134	138	138	145	141	140	87	923
Ethyl benzene (Y_{41})	136	142	142	146	142	142	88	938

Summary: Chemical complex optimization is an effective approach for economic improvement, source reduction, and sustainable development in a chemical production complex. The chemical production complex in the lower Mississippi River corridor has been used to show how those potentially new plants can be integrated into this existing infrastructure using the Chemical Complex Analysis System. The optimum configuration of plants was determined based on economic, environmental and sustainable costs using the System. The System was used for multi-criteria optimization to find optimal solutions that maximize companies' profits and minimize costs to society, called efficient or Pareto optimal solutions. It is another decision to determine the specific value of the weight that is acceptable to all concerned. Monte Carlo simulation was used to determine the sensitivity of the optimal solution to the costs and prices for the chemical production complex triple bottom line model with the cumulative probability distribution, a curve of the probability as a function of the triple bottom line. Now a range of values is available for the optimum triple bottom line that can be used to assess the risk of proceeding as measured by the cumulative probability distribution.

Multi-criteria optimization has been used with Monte Carlo simulation to determine the sensitivity of the optimal structure of a chemical production complex to prices, costs, and sustainable credits/cost. In essence, for each Pareto optimal solution, there is a cumulative probability distribution function that is the probability as a function of the triple bottom line. This information provides a quantitative assessment of the optimum profit versus sustainable credits/cost, and the risk (probability) that the triple bottom line will meet expectations.

The capabilities of the Chemical Complex Analysis System have been demonstrated by determining the optimal configuration of units based on economic, environmental and sustainable costs. Based on these results, the methodology could be applied to other chemical production complexes in the world for reduced emissions and energy savings.

References

Adler, S. F. 1999, Sustainability Metrics Interim Report No. 1 and Interim Report No. 2 AIChE/CWRT, 3 Park Avenue, New York, NY.

Anonymous, 1998, 1996 *Toxic Release Inventory, State Fact Sheets*, U. S. Environmental Protection Agency, Office of Pollution Prevention and Toxics (7408), Washington, D. C. (May, 1998).

Arnold, R. A., 2008, *Economics*, ISBN 0-324-59542-5, Cengage Learning, USA

Aspen Technology, 2013, *http://www.aspentech.com/products/aspen-kbase.aspx*, accessed 9-10-2013.

Banerjee, A., 2010, "The Advent of Corn-based Ethanol: A Re-examination of the Competition for Grains," RIS Policy Briefs, No. 46, March 2010

Beaver, E. and B. Beloff, 2000, Sustainability Metrics for Industry Workshop, AIChE/CWRT and BRIDGES to Sustainability, Austin, Texas, May 17-18, 2000.

Blau, G. E. and K. E. Kuenker, 1998, "Cultural Shift: Positioning Technical Computing to Enable Sustained Profitability in the Specialties Business," *Foundations of Computer Aided Process Operations*, AIChE Symposium Series, Vol. 94, No. 320, p 127.

Brown, T. R., 2000, "Estimating Product Costs," *Chemical Engineering*, August, 2000, p. 86-89.

Brundtland Report, 1987, Brundtland, G. H., 1987, "Our Common Future," Report of the World Commission on Environment and Development, United Nations, 247 pages.

Cabezas, H., J. C. Bare and S. K. Mallick, 1997, *Computers Chem Engr*, Vol. 21, Supp S305.

Cassinatis, P., 1988, *A Concise Introduction to Engineering Economics*, Union Hyman, Inc., Winchester, MA.

C&E News, 2009, "Chemical Output Slipped in Most Regions," *Chemical & Engineering News*, 87(27): 51

Chemical Engineering Magazine, 2011, Economic Indicators, p.58, August 2011, *www.che.com* accessed 9-18-2013

Constable, D. et al., 1999, Total Cost Assessment Methodology; Internal Managerial Decision Making Tool, AIChE/CWRT, AIChE, 3 Park Avenue, New York, NY, February 10, 2000.

Cussler, E. L., 1999 "Do Changes in the Chemical Industry Imply Changes in the Curriculum?" *Chemical Engineering Education,* Vol. 33, No. 1, p. 12-17.

Cussler, E. L. and G. D. Moggridge, 2001, *Chemical Product Design*, Cambridge University Press, Cambridge, England

DeGarmo, E. P., W. G. Sullivan and J. A. Bonadelli, 1988, *Engineering Economy*, 8th Ed., Macmillan, New York, NY.

DeSimone, D. L. and F. Popoff, 1997, *Ecoefficiency: The Business Link to Sustainable Development*, MIT Press Cambridge, MA.

Douglas, J. M., 1988, *Conceptual Design of Chemical Processes*, McGraw-Hill, New York, NY.

Feng, Y. and G. P. Rangaiah, 2011, *Chemical Engineering*, August, 201

Friedler, F., J. B. Varga and L. T. Fan, *Chem Eng Science*, Vol. 58, No. 11, p. 1755.

Garrett, D. E., 1989, *Chemical Engineering Economics*, Van Nostrand Reinhold, New York, NY.

Gizinski, G. H., 2000, "The Essentials of a Winning Marketing Strategy," *Chemical Engineering*, p. 12–124 (October, 2000).

Grant, E. L., W. G. Iresons and R. S. Leavenworth, 1982, *Principles of Engineering Economy*, 7th ed. John Wiley & Sons, New York, NY.

Hall, R. S. J. Matley and K. J. McNaughton, 1984, "Cost of Equipment-Data and Estimating Methods," *Modern Cost Engineering Methods and Data*, Vol. II J. Matley, Ed. McGraw-Hill, New York, NY.

Holland, F. A., F.A. Watson and J.K. Wilkinson, 1983, *Introduction to Process Economics.* 2nd ed. John Wiley & Sons, New York, NY.

ICIS, 2009, "Ethylene," *ICIS Chemical Business*, 276(15): 40

Ikoku, Chi U., 1985, *Economic Analysis and Investment Decisions*, John Wiley & Sons, New York, NY.

Indala, Sudheer, 2004, *Development and Integration of New Processes Consuming Carbon Dioxide in Multi-Plant Chemical Production Complexes*, M. S. Thesis, Louisiana State University, Baton Rouge, LA

Johnson, J., 1998, *Chemical & Engineering News*, p. 34, August 17, 1998.

Kelly, 2004, Formulating Production Planning Models, *Chemical Engineering Progress*, Vol. 100, No. 1, p. 43-50, American Institute of Chemical Engineers, NY, January, 2004.

Knopf, F. C., 2012, *Modeling, Analysis and Optimization of Process and Energy Systems*, John Wiley & Sons, Hoboken, NJ.

Kocis D. and I. Grossmann, 1989, *Computers Chem Engr*, Vol. 21, No. 7, p. 797-819.

Kohlbrand, H. K., 1998, Proceedings of Foundations of Computer Aided Process Operations Conference, Snowbird, Utah, July 5-10, 1998.

Soares, J. B., et al, 2006, "Alternative Depreciation Policies for Promoting Combined Heat and Power (CHP) Development in Brazil," Energy, Vol. 31, p. 1151-1166, Elsevier, New York

Kment, R., 2009, Commodities Report, *Ethanol Producers Magazine*, p. 21, May 2009, p.21 June 2009, p. 23, July 2009, September, 2009p. 19, October 2009, p.19, November 2009, p. 19, 2010

Koninckx, J., T. J. McAvoy, T. E. Martin, 1988, "On-Line Optimization Using Steady State Models," AIChE Annual Meeting, Washington, D. C.

Lindeberg, M. R., 1998, *Engineer-in-Training Reference Manual*, 8th Edition, Professional Publications, Belmont, CA

Luchasky, M.S. and Monk, J. 2009, "Supply and Demand Elasticities in the U. S. Ethanol Fuel Market," *Energy Economics*, Vol. 31, p. 403-410.

McConnell, Chia (Worley-Parsons), 2009, Comparison of Equipment and Labor Cost Indices, Source: Strategic Analysis of the Global Status of Carbon Capture and Storage, Report 2: Economic Assessment of the Carbon Capture and Storage Technology, Global CCS Institute, Canberra, Australia

Newnan, D. G., T. G. Eschenbach, J. P. Lavelle, 2004, *Engineering Economic Analysis*, Ninth Edition, Oxford University Press, Oxford, England

Perry's Chemical Engineers' Handbook, 1997, Seventh Edition, D. W. Green, Ed., McGraw-Hill, New York, NY.

Peters, M. and K. Timmerhaus, 1991, *Plant Design and Economics for Chemical Engineers*, 4th Ed., McGraw-Hill, New York, NY

Resnik, W., 1981, *Process Analysis for Chemical Engineers*, McGraw-Hill, New York, NY

Resnik, W., 1981, *Process Analysis for Chemical Engineers*, McGraw-Hill, New York, NY.
Thayer, A. M. 2001, "Financial Risks," *C&E News*, Feb. 26, 2001.

Sengupta, D. and R. W. Pike, *Chemicals from Biomass: Integrating Bioprocesses into Chemical Production Complexes for Sustainable Development*, CRC Press, Boca Raton, FL, 2012.

Sepulveda, J. A., W. E. Souder and B. S. Gottfried, 1984, *Schaum's Outline of Theory and Problems of Engineering Economics*, McGraw Hill, New York, NY.

Towler, Gavin and Ray Sinnott, 2013, *Chemical Engineering Design: Principles, Practice and Economics of Plant and Process Design*, Second Edition, Butterworth-Heinemann, Elsevier, Oxford, UK

Turton, R., R. C. Bailie, W. B. Whiting and J. A. Shaeiwitz, 2008, *Analysis, Synthesis and Design of Chemical Processes*, 1st Edition, Prentice Hall, Upper Saddle River, NJ.

Turton, R., R. C. Bailie, W. B. Whiting and J. A. Shaeiwitz, 2012, *Analysis, Synthesis and Design of Chemical Processes*, 4th Edition, Prentice Hall, Upper Saddle River, NJ.

Ulrich, G. D., 1984, *A Guide to Chemical Engineering Process Design and Economics*, John Wiley and Sons, New York, NY.

USDA, 2010, "Feed Grains Database: Yearbook Tables", http://www.ers.usda.gov/data/feedgrains/, accessed 8/4/2010

U. S. Master Tax Guide, 2013, CCH Editorial Staff Publication, CCH: A Woltes Kluwer Business, Chicago, IL

Vatavuk, W. M., 2002, "Updating the CE Plant Cost Index," *Chemical Engineering*, January 2002.

Valle-Riestra, J. F., 1983, *Project Evaluation in the Chemical Industry*, McGraw-Hill, New York, NY

Ward, T. J., 1986, "Profitability Analysis," *Plant Design and Cost Estimation*, Vol. 1, Modular Instruction Series, American Institute of Chemical Engineers, New York, NY.

Wikipedia, 2010, Elasticity (economics), http://en.wikipedia.org/wiki/Elasticity_%28economics%29 Accessed June 21, 2010

Xu, Amin, 2004, *Chemical Production Complex Optimization, Pollution Reduction and Sustainable Development*, PhD Dissertation, Louisiana State University, Baton Rouge, LA

Problems and Solutions

Interest Problems

1. What amount of money must be invested to earn $700 in simple interest in 8 years at 10%?

2. Compare the amount of simple interest earned on $1,000 at 10% for 8 years with the amount earned if the interest was compounded.

3. What amount of money would have to be invested to have $4,000 at the end of 3 years at a 10% compound interest rate?

4. If the inflation rate is 8%, and you invest $20,000 at an 11% simple interest rate, will you have retained your buying power at the end of: a. 5 years? B. 10 years?

5. How much would you deposit in an account to earn an annuity of $500 for 10 years if the account earns 12%?

6. How much would you deposit in an account to earn an annuity of $500 in the first year, increasing $50 each year to 10 years at 12%?

7. $500,000 is borrowed at a nominal rate of 8%, compounded quarterly. If no payments are made in the first 3 years, how much will be owed? How much would the amount be if interest was compounded annually? Daily?

8. If you are 25 years old now, and want to retire at 50 with an annuity account that would give you $5,000 a year for 30 years, how much do you need to deposit each month at 6% to be able to have enough in the account?

Investment Decision Problems

9. A company needs to decide whether to use some of their undistributed profits to automate certain of their procedures. The initial investment would cost $60,000 and the system would have a seven year lifetime. Savings from the automation are estimated at $18,000. Calculate the annual worth, the present worth, and the rate of return at 10%.

10. A manufacturer of motors for small household equipment, is wondering if they should close one of their two plants (Plant A) and expand operations at Plant B. It seems there is not enough business to keep both plants running at capacity, yet expansion of plant B would have to occur for it to handle double the usual load. Net income has been a steady $800,000 at plant A, and $840,000 at plant B. Due to savings in overhead costs, it is thought that, with the same level of business, net income from an upgraded plant B alone would equal $1,780,000. Salvage value of plant A and sale of the land is estimated at $120,000. The

investment to expand plant B would cost $1.1 million, and it would take three years before the level of production could be increased.

a. If the company has a required MARR of 11%, and operations are thought to be steady for the next 15 years, how desirable would this course of action be, if disassemble of plant A were to begin immediately and the total cost of upgrading was incurred at time zero?

b. If disassembling plant A was to occur at the end of year 3, when the expansion is completed?

c. If the disbursement for upgrading was spread evenly over the first three years of construction?

11. A corporation has $225,000 in its capital budget to invest this year. The following seven proposals for projects to reduce costs or raise profits are being considered.

Project	Investment Required	Estimated Life (Years)	Economic Salvage Value	Estimated Annual After Tax Cash Flow
1	$18,000	5	$ 2,000	$ 4,000
2	70,000	8	10,000	12,000
3	60,000	12	0	14,000
4	45,000	10	2,000	4,800
5	52,000	8	5,000	18,000
6	10,000	5	1,000	3,400
7	85,000	15	5,000	1,800

a) Calculate the return on investment for each project, and rank the projects accordingly. Which projects should be recommended, if no other factors are to be considered? What will the company's MARR be for these projects?

b) Suppose management is averse to the risk inherent in a long term and wants to consider profitability over only an 8-year period. How would the choice of projects and the MARR change?

12. A paper company is planning a new mill, but is uncertain of the capacity. The income increases with the capacity of the plant. However, the company's MARR of 17% must be met. This generates seven alternatives given below.

	1	2	3	4	5	6	7
Investment	$50,000	55,000	60,000	65,000	70,000	75,000	88,000
After tax cash flow	10,000	15,000	18,000	20,000	21,000	22,000	25,000

a. Compute the return on investment for each alternative.

b. Calculate the marginal return on each increment of investment.

c. Which alternative is the best?

13. A company is considering purchasing a piece of equipment to reduce costs, and there are five models to be evaluated. If their MARR is 20%, there is no salvage value and the equipment will last eight years, which model should be chosen?

Model	Investment	Cost Reduction (annual)
1	$ 82,500	$18,010
2	93,200	22,020
3	98,066	22,252
4	104,400	30,003
5	110,110	32,580
6	112,000	33,008

14. Evaluate the models in the previous problem using the following information. The tax rate is 33%; the equipment is depreciated over five years; and the company's after-tax MARR is 15%.

a. Evaluate the after-tax cash flows for each model and calculate the return on investment.

b. Determine the best model based on an incremental analysis of cash flows and return on investment.

15. A city is planning a new administrative building for additional space. Also, city planners anticipate a need for even more space in a few years. Therefore, a decision must be made about including the capability to add another floor to the building in the future. This capability would add $90,000 to the $720,000 estimated for construction of the building. A second story would cost $400,000 if the structural support is included and $540,000 if it is not. The building is to have a 40-year life, and maintenance costs are not affected by the decision. When must expansion occur to justify the additional expense using a MARR of 8%? Taxes are not considered since this is a government building.

16. Refer to Example 5 on present worth.

a. Consider that there is no salvage value, and compute the present worth. How sensitive is the new product to an estimate of salvage value?

b. Consider that maintenance of $40,000 is done every 5 years instead of every 10 years, excluding year 20. How sensitive is the new project to an estimate of maintenance?

c. Consider the after tax cash flow is decreased by $3,000. How sensitive is the new project to an estimate of the after tax cash flow?

d. Compute the break-even point, i.e. the value for the after-tax cash flow that the present worth is zero, and give an interpretation of this number.

17. Refer to Example 5 on present worth. There is a 50% probability that the annual profit will increase by $500. There is a 10% probability that the profit will stay at $6,000. There is a 15% probability that it will increase by $250, and a 25% probability it will increase by $700. Compute the expected value of the present worth.

18 ABS resin is produced by emulsion polymerization. The Process Economics Program of SRI International gives a breakdown of the various costs associated with three plant capacities in their PEP yearbook for this process. For a plant located in the Gulf coast, this information has been consolidated for a plant that produces 50,000 tons/yr (100 million pounds/yr) of ABS

Plant Costs		Product Costs	
Plant installed cost	$48.6 million	Raw materials	$55.2 million/yr
Other plant costs	$20.7 million	Utilities	$1.8 million/yr
Total plant cost	$69.3 million	Labor	$6.0 million/yr
		Other	$9.1 million/yr
		Total product cost	$72.1 million/yr

Estimated annual sales	$145 million/yr
Economic life	20 yrs.
Tax rate	38%
Minimum attractive rate of return	25% after taxes
Annual capital expenditures	$5.5 million/yr for worn out equipment

Depreciation: Straight-line method with salvage value of $10.3 million.

Perform the following economic analysis and determine:

a. Net annual income before taxes
b. Net annual cash flow before taxes
c. Depreciation
d. Taxes
e. Net annual income after taxes
f. Net annual cash flow after taxes
g. Net present value based on the net annual cash flow after taxes and minim attractive rate of return after taxes of 25%.
h. Show that 58% is the rate of return where the net present value is zero.
i. Determine the economic price based on a 58% rate of return.
j. Describe how the net present value, rate of return and economic price are used for economic decision.

19. Show that the economic price has the net present value equal to zero, i.e., derive the economic price equation starting with the equation for the Net Present Value, Equation (24) equal to zero.

20. Normal butane is isomerized to more valuable isobutane in a catalytic reactor. The activity of the catalyst declines with time and the catalyst has to be regenerated. The cost to regenerate the catalyst is $40,000, and it returns the plant to making an annual net profit of $135,000. After two years the profit has dropped to $120,000/yr, and after three and four years the profit is $105,000/yr and $85,000/yr.

a. Convert the cost to regenerate the catalyst, $40,000, to an equivalent uniform annual cost with an interest rate of 25% and an economic life of 2.0 years.

b. Compare the profit including the cost of regeneration with the cost of continuing to operate at years two, three and four.

c. Determine the year to regenerate the catalyst to have the maximum annual profit.

21. A refinery plans to add an alkylation process to convert butenes and isobutane to isooctane for gasoline blending. Correlations are given below to estimate the total plant cost and total product cost as a function of the isooctane product flow rate (plant capacity), m, in thousand bbls/day. (Brikler, *Chemical Engineering*, p. 129, Sept. 1987).

Total plant cost ($ million) $C_{total\ plant} = 12.0m^{1.2}$
Total product cost ($million/yr) $C_T = 0.1m^{1.2}$

Sales price of isooctane is $24.66/bbl for blending in gasoline and sales are given in terms of the product flow rate m in thousand bbls/day:

S($million/yr) = ($24.66/bbl)(1000bbl/thous.bbls)(m thous. bbls/day)(365days/yr)
                  ($ million/10^6)
or
Sales ($ million/yr) S = 9.0 m

A minimum attractive rate of return is 25%, and the economic live of the plant is 20 years.

a. Develop the equation for the Net Present Value for the proposed new alkylation process based on the annual cash flow before taxes, i.e., sales minus total product cost and not including annual equipment purchases,

b. The upper limit on the capacity, m, of the process is 500 thousand bbls/day. Give the optimization problem to be solved as:

maximize NPV = equation from part a.
subject to: inequality constraint on m, less than 500.c.

Solve the optimization problem in part b using the Kuhn-Tucker conditions and Lagrange multipliers, and determine the optimal values of m and NPV. There will be two cases.

Determine if these cases are maximum or minimum using the Kuhn-Tucker necessary conditions.

c. It is projected that the annual net profit for the alkylation process will be the following for the first five years to the first major turn-around. The capital investment cost is $66 million for a capacity of 12,000 bbl per day.

End of Year	Annual Net Profit ($ million)
1	32.0
2	28.0
3	22.0
4	17.0
5	10.3

i. Compute the net present value, NPV, for interest rates of 15% and 25% for the alkylation unit. Give your conclusion about this proposed investment.

ii. Determine the rate of return for the alkylation unit.

iii. Calculate the payback period. Why should this method <u>not</u> be used in engineering economics calculations?

22. The plant design shown in the process flow diagram will produce 3200 tons per day of 93% sulfuric acid. The contact process uses three steps to produce sulfuric acid and steam from air, molten sulfur and water. First, molten sulfur feed is combusted with dry air in the sulfur burner, $S + O_2 \Rightarrow SO_2$. In the four packed bed reactors, $SO_2, + \frac{1}{2}O_2 \Rightarrow SO_3$. In the absorber section, SO_3 is absorbed by 98 wt% sulfuric acid to produce a more concentrated acid by the reaction, $SO_3 + H_2O \Rightarrow H_2SO_4$. The exit gases from the final absorption tower are discharged to the air with less than 4 lb. of SO_2 per ton of sulfuric acid produced.

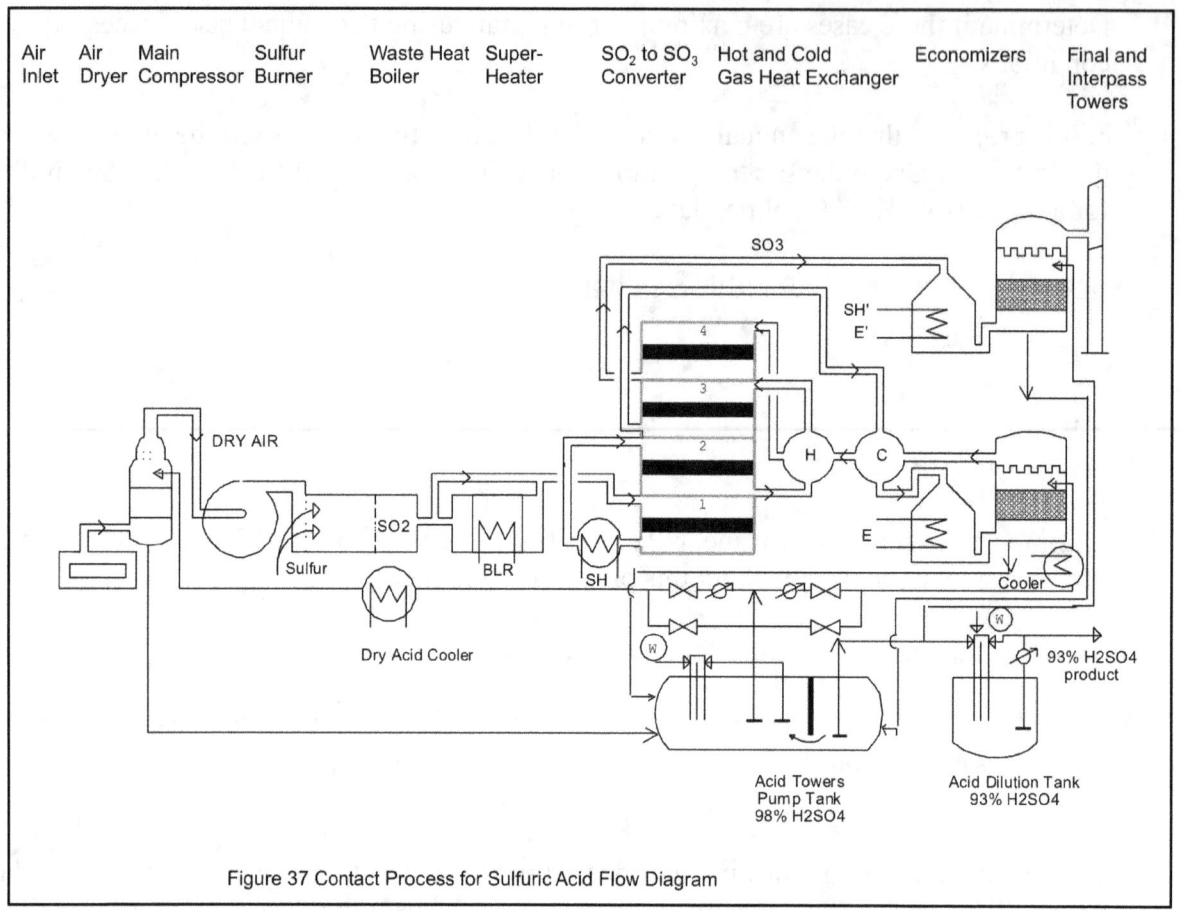

Figure 37 Contact Process for Sulfuric Acid Flow Diagram

The following information was developed for the economic evaluation for the plant design of the contact process.

Plant Costs		Product Costs	
Plant installed cost	$23.2 million	Raw materials	$7.3 million/yr
Other plant costs	$10.2 million	Utilities	$1.8 million/yr
Total plant cost	$33.4 million	Labor	$3.2 million/yr
		Other	$2.1 million/yr
		Total product cost	$14.4 million/yr

Estimated annual sales (sulfuric acid at $25/ton and steam) $58.4 million/yr
Economic life of the plant   30 yrs.
Tax rate   38%
Minimum attractive rate of return   25% after taxes
Annual capital expenditures $1.2 million/yr for worn out equipment
Depreciation: Straight-line method with salvage value of $4.8 million.

128

Perform the following economic analysis and determine:

a. Net annual income before taxes
b. Net annual cash flow before taxes
c. Depreciation
d. Taxes
e. Net annual income after taxes
f. Net annual cash flow after taxes
g. Net present value based on the net annual cash flow after taxes and minim attractive rate of return after taxes of 25%.
h. Rate of return.
i. Economic price for sulfuric acid in $per ton.
j. Describe how the net present value, rate of return and economic price are used for economic decisions.

23. Two cash flow diagrams in Figure 38 show two proposed schedules for the cost of maintaining a chemical reactor over a four-year period. The catalyst in the reactor degrades over time. Also, the pressure drop increases because the catalyst pellets crumble, and the catalyst has to be screened to remove fine particles. In addition, gratings and interior supports erode and corrode, and they have to be replaced. Compute the net present value of these cash flows using a MARR of 15% and determine which schedule is better. Arrows are not to scale.

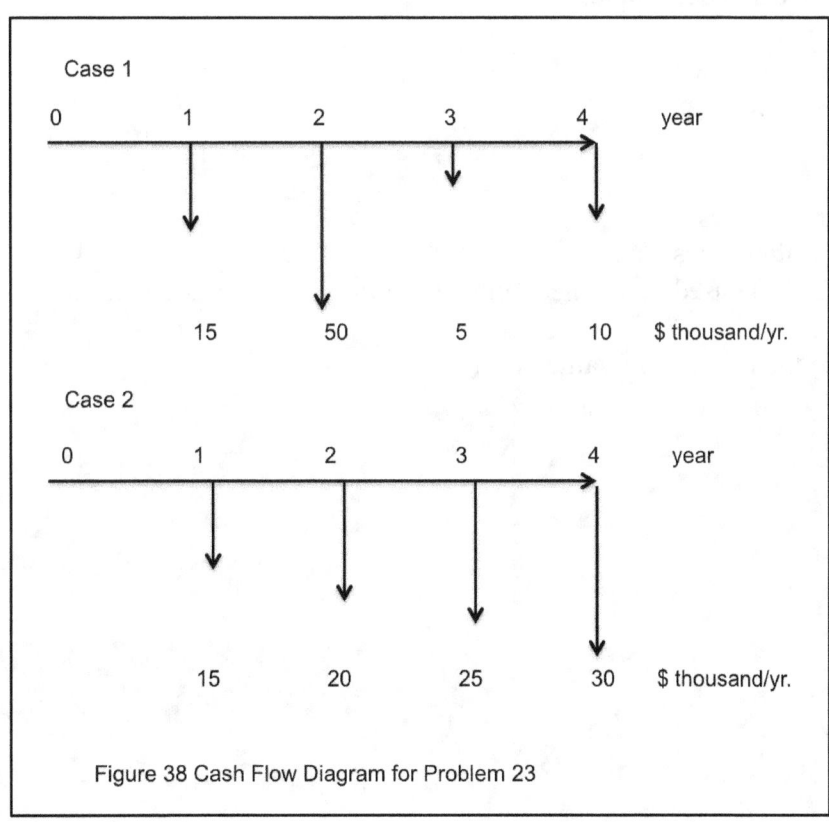

Figure 38 Cash Flow Diagram for Problem 23

24.   Refer to Example 5 on present worth. There is a 50% probability that the annual profit will increase by $500. There is a 10% probability that the profit will stay at $6,000. There is a 15% probability that it will increase by $250, and a 25% probability it will increase by $700. Compute the expected value of the present worth.

25.   Perform an economic analysis for a monochlorobenzene separation process using the following cost data has been obtained for the product purification facilities of a monochlorobenzene (MCB) process. This part of the process purifies the product from the reactor that is used to produce MCB by the chlorination of benzene. (Seider, et al., 2004)

Plant capacity	44.1 million pounds/yr.of MCB
Plant installed cost	$1.0 million
Total plant cost	$4.0 million
Total product cost	$19.3 million/yr
Annual capital expenditures for worn out equipment	$0.1 million/yr
Estimated annual sales	$21.4 million/yr
Economic life	15 yrs
Tax rate	30%
Minimum attractive rate of return	15%

Depreciation: straight-line method with no salvage value.

Use the information above and determine:

a. Net annual income before taxes
b. Net annual cash flow before taxes
c. Depreciation
d. Taxes
e. Net annual income after taxes
f. Net annual cash flow after taxes
g.   Net present value (NPV) based on the net annual cash flow after taxes and minim attractive rate of return.
h. Rate of return where the net present value is zero.
i. Economic price.

26. Economic Analysis for an Acetone Plant Design. The following cost data has been obtained from the design of a plant to produce acetone from isopropyl alcohol.

Plant capacity (product flow rate) 15.8 million kg/yr.of acetone
Estimated annual sales $28.0 million/yr
Plant installed cost $1.69 million
Total plant cost $2.61 million
Total product cost $26.5 million/yr
Annual capital expenditures $0.05 million/yr
 for worn out equipment
Economic life   10 yrs
Tax rate   35%
Minimum attractive rate of return   15%
Depreciation: straight-line method with no salvage value.

Use the information above and determine:
a. Net annual income before taxes
b. Net annual cash flow before taxes
c. Depreciation
d. Taxes
e. Net annual income after taxes
f. Net annual cash flow after taxes
g. Net present value (NPV) based on the net annual cash flow after taxes and minim attractive rate of return.
h. Rate of return where the net present value is zero.
i. Economic price in $/kg.

27. Paraffins are reformed to more valuable aromatics in a catalytic reactor. The activity of the catalyst declines with time and the catalyst has to be regenerated. The annual profit is given by:

Year	1	2	3	4	5
Annual net profit	$135,000	$120,000/yr	$105,000/yr	$85,000/yr	$65,000

The cost to regenerate the catalyst is $40,000, and regeneration returns the plant to making an annual net profit of $135,000.

a. Convert the cost to regenerate the catalyst, $40,000, to an equivalent uniform annual cost with an interest rate of 25% and an economic life of 2.0 years.

b. Compute the annual profit for each year of the five years if the catalyst is regenerated at the start of year 2 and the equivalent uniform annual cost are charged to years 2 and 3, i.e. year 2, $135,000 - EUAC and year 3, $120,000 - EUAC.

c. Compute the annual profit for each year of the five years if the catalyst is regenerated at the start of year 3 and the equivalent uniform annual cost are charged to years 3 and 4, i.e. year 3, $135,000 - EUAC and year 4, $120,000 - EUAC.

d. Compute the annual profits for each year of the five years if the catalyst is regenerated at the start of year 4 and the equivalent uniform annual cost are charged to years 4 and 5, i.e. year 4, $135,000 - EUAC and year 5, $120,000 - EUAC

e. Determine the year to regenerate the catalyst to have the maximum annual profit for the five years.

28. Perform an economic evaluation for the acetone from isopropyl alcohol plant design using the following description of a 1.88 tonne/hr plant for acetone from 2.40 tonne/hr of isopropyl alcohol from Turton, et al., 1998. See Turton, et al., 1998 for the process flow diagram (PFD), a process description, equipment and utilities from the plant design, and cost of some chemicals.

The process for dehydrogenating isopropyl alcohol to produce acetone has the advantage that the acetone is free of aromatic impurities as compared being manufactured from phenol. In the reactor the conditions are 2.0 bar and 350°C with a single pass conversion of 90%. At the start of the process, a flow rate of 51.96 kmol/hr of an azeotropic mixture of isopropyl alcohol and water (67.0% mol isopropyl alcohol) is mixed with flow rate of 5.88 kmol/hr of a recycle stream of unreacted isopropyl alcohol (56.8 % mol isopropyl alcohol, 2.4 % mol acetone, remaining water). This stream is vaporized and sent to a reactor where heat is provided for the endothermic reaction. The products from the reactor include 34.94 kmol/hr of acetone, 34.78 kmol/hr of hydrogen, 3.86 kmol/hr of isopropyl alcohol, and 19.04 kmol/hr of water, and it is cooled with two heat exchangers before entering the phase separator. The vapor leaving the separator contains 4.44 kmol/hr of acetone, 34.78 kmol/hr of hydrogen, 0.12 kmol/hr of isopropyl alcohol, and 0.40 kmol/hr of water, which goes to the acetone stripper column to recover acetone. The liquid from the acetone stripper column is combined with the liquid from the phase separator to have a stream that contains 32.43 kmol/hr of acetone, 0.00 kmol/hr of hydrogen, 3.84 kmol/hr of isopropyl alcohol, and 37.75 kmol/hr of water. This stream is sent to two distillation towers to have a 99.9 % mol acetone product (32.27 kmol/hr of acetone, 0.02 kmol/hr of isopropyl alcohol), and a recycle stream of 0.16 kmol/hr of acetone, 3.82 kmol/hr of isopropyl alcohol and 1.90 kmol/hr of water.

The catalytic reaction of isopropyl alcohol to form acetone is kinetically controlled and is first order with respect to the concentration of alcohol. A rate equation and rate constants are reported by Turton, et al., 1998 who also said that several side reactions occur only to a small extent. The vapor-liquid phase equilibrium was simulated using UNIFAC in the CHEMCAD simulator.

a. Develop an Excel workbook with three spreadsheets to perform the economic evaluation of the plant. The program CAPCOST* or other comparable program will be used to estimate the cost of the major process equipment given in the PFD.

Do the following calculations using an Excel workbook named *Economic evaluation for acetone process.xls* and CAPCOST.

b. Determine the installed costs of the equipment on the PFD using program CAPCOST. This program evaluates cost based on 1996 Chemical Engineering Plant Cost Index (CEPCI) of 382. The 2001 value is 394. For the purposes of this evaluation, consider CAPCOST's bare module cost as installed cost. For the reactor R-401 use a fixed tube sheet S&T Exchanger with an area of 468 m², and for the reactor furnace H-401 use a fired heater, thermal fluid heater - molten salt with a duty of 758.3 kW. If a satisfactory value for a piece of equipment is not available from CAPCOST, use the figures from AppendixA. Put these installed costs in the spreadsheet *Plant*.

c. Evaluate the total plant cost (total capital investment) using the spreadsheet *Plant*

d. Estimate the total product cost on an annual basis using the spreadsheet *Product*.

e. Estimate the annual sales on an annual basis using the spreadsheet *Product*

f. Perform the economic analysis shown in a spreadsheet *EcoAna*.

i. Evaluate the net present value for a MARR of 15% and an economic life of 10 years.

ii. Evaluate the rate of return for an economic life of 10 years.

iii. Evaluate the economic price for a 20% return on capital employed and an economic life of 10 years.

*CAPCOST is a cost estimation program from the book by R. Turton, R. C. Bailey, W. B. Whiting and J. A. Schaeiwitz, *Analysis, Synthesis and Design of Chemical Processes*, Prentice-Hall, Upper Saddle River, NJ, 1998 where a detailed description of the program is given.

Solutions

1. $I = n \cdot i \cdot P$ or $\$700 = 8 \cdot 0.10 \cdot P$ or $P = \$875$

2. (simple interest) $I = n\,i\,P$ or $I = 8 \cdot 0.10 \cdot 1000 = \$800$
   (compound interest) $I = F - P = P(1 + i)^n - P = \$1,000(1 = 0.10)^8 - 1000 = \$1,143$

3. $P = \dfrac{F}{(1 + i)^n} = \dfrac{4000}{(1 + 0.10)^3} = = \$3,005$

4. a. The principal will grow to: $F = \$20,000[1 + 5(0.11)] = \$31,000$
To see if the buying power is the same, the future amount must be discounted for inflation, so that $\$1.00$ in the future can be converted to an equivalent amount today.
$$F_I = \frac{P_I}{(1 + r)^n}$$
where F_I is the future worth, measured in today's dollars P_I, and where r is equal to the rate of inflation. Notice that discounting for inflation is the reverse of compound interest problems.
$$F_I = \frac{31,000}{(1 + 0.08)^5} \qquad F_I = \$21,098 \quad \text{Buying power is retained, plus a little extra.}$$

b. $F = \$20,000[1 + 10(11)] = \$42,000$

$$F_I = \frac{42,000}{(1 + 0.08)^{10}} = \$19,454 < \$20,000$$

With an increase in time, buying power diminishes because of the compounding nature of inflation.

5. P = \$500(P/A, 12%, 10) = \$2,825

6. P = \$500(P/A, 12%, 10) + 50(P/G, 12%, 10) = 2825 + 1013 = \$3838

7. For quarterly compounding, first fund the effective interest rate from Equation 3.
   $(1 + 0.08/4)^4 - 1 = 8.24\%$

F = \$500,000 (F/P, 8.24%, 3) = \$636,064

Compounded annually
F = \$500,000(F/P, 8%, 3) = \$629,856

Compounded daily
F = \$500,000(F/P, 8/365%, 1095) = \$635,608

8. First, find the present value of the annuity:

   P = \$5,000(P/A, 6%, 30) = \$68,824

This is the future target amount. To get the annual deposit required:
A = \$68,824(A/P, 6%, 25) = \$5,384

9. PW = \$27,631, AW = \$5,676, ROR = 22.92%

10. The first task is to separate all factors unique to the proposed course of action. If disassembling the plant were to occur immediately, that is, at time zero, a cost of \$980,000 would be incurred, which is the expenditure for upgrading minus the income from sale and salvage of Plant A. The first three years of the project would incur a loss of \$800,000 from Plant A, with no changes in income from Plant B. The next fifteen years would see a positive cash flow of \$140,000, which is the change in income caused by the project.

a. If the company has a required MARR of 11%, and operations are thought to be steady for the next 15 years, how desirable would this course of action be, if disassemble of plant A were to begin immediately and the total cost of upgrading was incurred at time zero?

The internal rate of return is a negative 4.65%, and the return on investment is -\$2,198,866, clearly unacceptable.

b. If disassembling is delayed until the third year, the cost in year zero is \$1.1, no income change occurs in years one and two, with a positive cash flow in year 3 from sale of land and

Plant A. Income of $140,000 is then constant for 15 years. Delaying the plant disassembly is clearly the more practical choice, yet return on investment at 7.4%, still does not meet the requirement of the company. Net present value is -$276,150.

c. If the disbursements for upgrading were spread evenly over the three years, instead of a lump sum at time zero, then at the close of years one and two, -$366,667 is incurred, -246,667 at year 3 (366,667 - 120,000), and $140,000 for the fifteen years thereafter. This makes the investment more attractive at a return on investment of 9.6%, but this is still not sufficient to meet an MARR of 11%. The net present value is -$72,180.

11. a. The order of the projects would be: 5, 6, 3, 7. The MARR corresponds to the rate of return of the last project chosen, i.e. 19.9%.

b. The order of the projects becomes 5, 6, 3, and 4. The MARR is 14.3%, and $58,000 is available for a new project if it can be found at this MARR.

12. Year	1	2	3	4	5	6	7
a. Return on Investment	5.47	16.19	19.91	20.93	19.91	19.01	17.75
b. Marginal ROI	98.36		55.81	32.66	5.47	5.47	8.16

c. Alternative four is the best because the marginal return on investment for alternative five is less than 17%. None of the larger capacities have a marginal ROI of 17% or greater.

13. ROR	Increment of Investment	Increment of Cost Reduction	Before Tax ROR on Increment of Investment
1 14.38%			
2 16.81%			
3 15.55%			
4 23.39%			
5 24.44%	$5,710	$2,577	42.47%
6 24.3%	$1,890	$428	15.49%

Note: 110,110 - 104,400 = 5,710, 113,000 - 110,110 = 1,890, 32,580 - 30,003 = 2,577, 33,000 - 32,580 = 428

Models 1, 2, and 3 do not meet the MARR, and it is not necessary to calculate their marginal return on investment. Comparing Models 4 and 5, Model 5 has the largest ROR and turns out to be the best one. The marginal return on investment of Model 6 is smaller than Model 5.

14.a.	Model 1 ATCF	Model 2 ATCF	Model 3 ATCF
Year			
1	$17,512	$20,905	$23,391
2	20,779	24,595	27,275

3	18,601	22,368	24,686
4	17,579	19,674	22,097
5	14,245	17,214	19,508
6	12,067	14,753	16,919
7	12,067	14,753	16,919
8	12,067	14,753	16,919

ROR = 11.48%   ROR = 13.13%   ROR = 15.23%

	Model 4 ATCF	Model 5 ATCF	Model 6 ATCF
Year			
1	$26,992	$29,096	$29,507
2	31,127	33,456	33,943
3	25,056	30,549	30,986
4	25,614	27,643	28,029
5	22,858	24,736	25,072
6	20,102	21,829	22,115
7	20,102	21,829	22,115
8	20,102	21,829	22,115

ROR = 17.47%   ROR = 18.94%   ROR = 18.85%

b)	Comparison of Models 3 and 4	Comparison of Models 4 and 5	Comparison of Models 5 and 6
Year			
1	$3,601	$2,104	$411
2	3,852	2,329	487
3	370	5,493	437
4	3,517	2,029	386
5	3,350	1,878	336
6	3,183	1,727	286
7	3,183	1,727	286
8	3,183	1,727	286

Increment of investment	$6,334	$5,710	$1,890

After-tax Incremental ROR	46.34% ROR =	43.18%	12.03%

Model 5 is the best choice from this after-tax analysis.

15.   5.75 years

16.   a.   P.W. = $4,770

       b.   P.W. = $3,618

       c.   P.W. = $4,459

       d. $0 = -150{,}000 + X(P/A, 10\%, 20) - \$40{,}000(P/F, 10\%, 10)+$
          $\$30{,}000(P/F, 10\%, 20) = \$160{,}963 + X(8.513)$ $X = \$18{,}908$

If the after-tax income is greater than $18,908, all other things being equal, then the project is a desirable one.

17. $0.5(\$6{,}500) + 0.10(\$6{,}000) + 0.15(\$6{,}250) + 0.25(6{,}700) = \$6{,}463$

18.   Perform the following economic analysis and determine:

a.   Net annual income before taxes = sales - total product cost

      = $145million/yr - $72.1 million/yr = $72.9 million/yr

b.   Net annual cash flow before taxes

      = net annual income before taxes - annual capital expenditures for worn out equipment

      = $72.9 - $5.5 = $67.4 million/yr

c.   Depreciation = (plant installed cost - salvage value)/economic life

      = ($48.6 million - $10.3 million)/20yrs. = $1.92 million/yr.

d.   Taxes = tax rate*taxable income = tax rate*(net annual income before taxes- depreciation)

      = 0.38($72.9 million/yr - $1.92 million/yr) = $27.0 million/yr

e.   Net annual income after taxes = net annual income before taxes - taxes

      = $72.9 - $27.0 = $45.9 million/yr

f.   Net annual cash flow after taxes

      = net annual income after taxes - annual capital expenditures for worn out equipment

      = $45.9 - $5.5 = $40.4 million/yr

g. Net present value (NPV) based on the net annual cash flow after taxes and minim attractive rate of return after taxes.

CF_0 = total plant cost = $69.3 million

A = net annual cash flow after taxes = $40.4 million/yr

i = minimum attractive rate of return after taxes = 25%

n = economic life = 20 yrs.

$$NPV = -CF_0 + A\{[(1 - (1+i)^{-n}]/i\}$$
$$= -\$69.3 \text{ million} + \$40.4 \text{ million/yr}\{[1 - (1.25)^{-20}]/0.25\} - \$69.3 \text{ million}$$
$$+ \$157.7 \text{ million} = \$90.4 \text{ million}$$

h. Show that 58% is the rate of return where the net present value is zero

$$NPV = -CF_0 + A\{[(1 - (1+i)^{-n}]/i\}$$
$$= -\$69.3 \text{ million} + \$40.4 \text{ million/yr}\{[1 - (1.58)^{-20}]/0.58\}$$
$$= -\$69.3 \text{ million} + 69.6 \text{ million} = 0.3 \text{ million}$$

i. Determine the economic price based on a 58% rate of return.

$P = CF_0$ = total plant cost = $69.3 million
i = rate of return = 0.58
n = economic life = 20 yrs.
Annual cost of capital $= EUAC = P*(A/P) = P*\{i/[1 - (1+i)^{-n}]\}$

$$= \$69.3\{0.58/[1 - (1.58)^{-20}]\} = \$40.2 \text{ million/yr.}$$

Economic price = (total product cost + annual cost of capital + annual capital expenditures)
                 /product rate
= ($72.1 million/yr + $40.2 million/yr + 5.5 million/yr)/100 millions pounds/yr = $1.18/lb

Current price of ABS resin is $1.18 – 1.20/lb.

j. Describe how the net present value, rate of return and economic price are used for economic decisions.

Net present value gives the present value of cash flows from a potential investment and has to be positive to be considered a viable project. The net present value of potential projects can be added to get a total net present value.

Rate of return is the interest rate for a zero net present value. It is used to compare returns from unlike investments such as new plants, bonds, real estate, and other types of business ventures.

Economic price gives the price that if the product is sold at this price will give the return on investment over the economic life of the plant. It is not related to the market price.

19. To show that the economic price has the net present value equal to zero, i.e., derive the economic price equation starting with the equation for the Net Present Value, Equation 24 equal to zero.

$$NPV = -CF_o + A[1 - (1+i)^{-n}]/i = 0$$

where A is the net annual cash flow and is equal to revenue from sales minus total product cost, C_T.

Revenue from sales is the product of sales price and product flow rate, i.e., $(S_p)(m)$, and CF_o is the total plant cost, $C_{total\ plant}$. For a zero NPV the above equation can be written:

$$0 = -C_{total\ plant} + [(S_p)(m) - C_T - C_{cap}][1 - (1+i)^{-n}]/i$$

Rearranging gives:

$$C_{total\ plant}/[1 - (1+i)^{-n}]/i = (S_p)(m) - C_T - C_{cap}$$

or

$$S_p = \{C_T + C_{total\ plant}[i/[1 - (1+i)^{-n} + C_{cap}]\}/m$$

which is the same as the definition of the economic price given by Equation 28 where the second term in the numerator is the annual cost of capital..

20 a. Equivalent uniform annual series, $EUAC = PW\{i/[(1 - (1+i)^{-n}]\}$, Equation 22, where PW = $40,000, cost to regenerate the catalyst, negative; i = 0.25 and n = 2 years.

$$EUAC = \$40,000\{0.25/[1 - (1.25)^{-2}]\} = \$27,800$$

Net profit from regenerating catalyst = $135,000 - $27,800 = $107,200

part b Comparison for years two, three and four

Year 2   Profit from continuing to operate (not regenerating catalyst) = $120,000
         Profit from operating with a regenerated catalyst   = $107,200

Year 3   Profit from continuing to operate (not regenerating catalyst) = $105,000
         Profit from operating with a regenerated catalyst   = $107,200

Year 4   Profit from continuing to operate (not regenerating catalyst) = $85,000
         Profit from operating with a regenerated catalyst   = $107,200

c. Use regenerated catalyst in year 3 to have the maximum annual profit of $107,200, since the profit from using a regenerated catalyst, $107,200 is greater than the profit for continuing to operate without regenerating the catalyst, $105,000. In year 2, the maximum annual profit is made by continuing to operate without regenerating the catalyst, $120,000.

21. part a, m is plant capacity

Total plant cost ($ million)	$C_{total\ plant} = 12.0m^{1.2}$
Total product cost ($million/yr)	$C_T = 0.1m^{1.2}$
Sales ($ million/yr)	$S = 9.0\ m$

$NPV = -CF_o + A\{[(1 - (1+i)-n/i]\}$ Equation 24

$CF_o = C_{total\ plant} = 12.0m^{1.2}$,   Annual cash flow before taxes $= A = S - C_T = 9.0\ m\ -\ 0.1m^{1.2}$

Substituting gives:

$NPV = -12.0m^{1.2}\ + (9.0\ m\ -\ 0.1m^{1.2})\{[(1 - (1+i)-n/i]\}$

for i = 0.25 and n = 20, $\{[(1 - (1+i)-n/i]\} = 3.95$

and

$NPV =\ -12.0m^{1.2}\ + (9.0\ m\ -\ 0.1m^{1.2})(3.95)$

simplifying gives:

$NPV = -12.395m^{1.2}\ + 35.55m$

part b

maximize: $-12.395m^{1.2}\ + 35.55m$
subject to: $m \le 500$

part c

Lagrange function:

$L(m, \lambda) = -12.395m^{1.2}\ + 35.55m + \lambda(\ m + x_s^2 - 500)$

$\dfrac{\partial L}{\partial m} =\ -12.395(1.2)m^{0.2} + 35.55 + \lambda = 0$

$\dfrac{\partial L}{\partial \lambda} =\ m + x_s^2 - 500 = 0$

$$\frac{\partial L}{\partial x_s} = 2 \lambda x_s = 0$$

Two Cases

Case I $\lambda = 0$, x_s not equal to 0

$-12.395(1.2)m^{0.2} + 35.55 = 0$

solving;

$m = (35.55/12.395 \cdot 1.2)^5 = 78.0$ thousand bbls/day

$x_s^2 = 500 - m = 422$

$NPV = -12.395(78.0)^{1.2} + 35.55(78.0) = \$462$ million

$\lambda = 0$, necessary condition for a maximum

Case 2: λ not equal to 0, $x_s = 0$

solving

$m = 500$

$\lambda = 12.395(1.2)(500)^{0.2} - 35.55 = 16.0$

$NPV = -12.395(500)^{1.2} + 35.55(500) = -3,704$

λ is positive, a necessary condition for a minimum.

c. It is projected that the annual net profit for the alkylation process will be the following for the first five years to the first major turn-around. The capital investment cost is $66 million for a capacity of 12,000 bbl per day.

End of Year	Annual Net Profit ($ million)
1	32.0
2	28.0
3	22.0
4	17.0
5	10.3

i. Compute the net present value, NPV, for interest rates of 15% and 25% for the alkylation unit. Give your conclusion about this proposed investment. Equation 23 gives the Net Present Value.

$$n$$

$$NPV = -CF_o + \sum_{j=1} CF_j(1+i)^{-j} \quad \text{where } (1+i)^{-j} = P/F \tag{23}$$

The following table computes the NPV for interest rates of 15% and 25%.

n	CF_j	P/F(15%)	CF_j (P/F)	P/F(25%)	CF_j (P/F)
1	32.0	1.150	27.83	1.250	25.60
2	28.0	1.323	21.17	1.563	17.92
3	22.0	1.521	14.47	1.953	11.26
4	17.0	1.749	9.72	2.441	6.96
5	10.3	2.011	5.12	3.052	3.38
			78.31		65.12

NPV (15%) = -66.00 + 78.31 = $12.31 million

NPV (25%) = -66.00 + 65.12 = - $0.88 million

A good investment if money is available at 15% interest rate.

Money is lost at 25% interest rate.

ii. Determine the rate of return (ROR) for the alkylation unit. Locate the interest rate where the NPV is zero.

ROR = 15% + [12.3/(12.3 – (-.88))] 10% = 24.33%

iii. Calculate the payback period (PBP). Why should this method <u>not</u> be used in engineering economics calculations? The PBP is given by Equation 26.

$$CF_o = \sum_{j=1}^{PBP} CF_j \tag{26}$$

66.0 = 32.0 + 28.0 + 22 x   where x = 0.27 and PBP = 2.27 years

This method should not be used because it ignores the time value of money.

22. a. Net annual income before taxes = sales - total product cost
    = $58.4 million/yr - $14.4 million/yr
    = $44.0 million/yr

b. Net cash flow before taxes = net annual income before taxes - annual capital expenditures for worn out equipment

    = $44.0 million/yr - $1.2 million/yr = $42.8 million/yr

c. Depreciation = (plant installed cost - salvage value)/economic life

$\quad$ = (\$23.2 million $-$ \$4.8 million)/30yrs. = \$0.61 million/yr.

d. Taxes = tax rate*taxable income = tax rate*(net annual income before taxes- depreciation)

$\quad$ = 0.38(\$44.0 million/yr - \$0.61 million/yr) = \$16.5 million/yr

e. Net annual income after taxes = net annual income before taxes $-$ taxes

$\quad$ = \$44.0 - \$16.5 = \$27.5 million/yr

f. Net annual cash flow after taxes

= net annual income after taxes - annual capital expenditures for worn out equipment

$\quad$ = \$27.5 - \$1.2 = \$26.3 million/yr

g. Net present value (NPV) based on the net annual cash flow after taxes and minim attractive rate of return after taxes of 25%.

$\quad$ CF_0 = total plant cost = \$33.4 million

$\quad$ A = net annual cash flow after taxes = \$26.3 million/yr

$\quad$ i = minimum attractive rate of return after taxes = 25%

$\quad$ n = economic life = 30 yrs.

$$\begin{aligned} \text{NPV} &= -CF_0 + A\{[(1 - (1+i)^{-n}]/i\} \\ &= -\$33.4 \text{ million} + \$26.3 \text{ million/yr}\{[1 - (1.25)^{-30}]/0.25\} \\ &= -\$33.4 \text{ million} + \$105.1 \text{ million} = \$71.7 \text{ million} \end{aligned}$$

h. Rate of return.

$\quad$ $\text{NPV}(i = 0.60) = -CF_0 + A\{[(1 - (1+i)^{-n}]/i\}$

$$\begin{aligned} &= -\$33.4 \text{ million} + \$26.3 \text{ million/yr}\{[1 - (1.60)^{-30}]/0.60\} \\ &= -\$33.4 \text{ million} + \$43.8 \text{ million} = \$10.4 \text{ million} \end{aligned}$$

$\quad$ $\text{NPV}(i = 0.70) = -CF_0 + A\{[(1 - (1+i)^{-n}]/i\}$

$$\begin{aligned} &= -\$33.4 \text{ million} + \$26.3 \text{ million/yr}\{[1 - (1.70)^{-30}]/0.70\} \\ &= -\$33.4 \text{ million} + \$37.6 \text{ million} = \$4.2 \text{ million} \end{aligned}$$

$$\text{NPV}(i = 0.80) = -CF_0 + A\{[(1 - (1+i)^{-n}]/i\}$$

$$= -\$33.4 \text{ million} + \$26.3 \text{ million/yr}\{[1 - (1.80)^{-30}]/0.80\}$$
$$= -\$33.4 \text{ million} + \$32.9 \text{ million} = -\$0.5 \text{ million}$$

Interpolating: $\text{ROR} = 0.70 + 4.2[(0.8 - 0.7)/(4.2 - (-0.5)] = 0.79$

i. Economic price for sulfuric acid.

$P = CF_0 = \text{total plant cost} = \$33.4 \text{ million}$
$i = \text{rate of return} = 0.25$
$n = \text{economic life} = 30 \text{ yrs.}$

Annual cost of capital $= \text{EUAC} = P*(A/P) = P*\{i/[1 - (1+i)^{-n}]\}$
$= \$33.4\{0.25/[1 - (1.25)^{-30}]\} = \$8.4 \text{ million/yr.}$

Economic price
$= (\text{total product cost} + \text{annual capital expenditures} + \text{annual cost of capital})/ \text{product rate}$
$= (\$14.4 \text{ million/yr} + \$1.2 \text{ million/yr} + 8.4 \text{ million/yr})/(3.200 \text{ tons/day})(365 \text{ days/yr})$
$= (\$24.0 \text{ million/yr})/(1,168,000 \text{ tons/yr}) = \$20.55/ \text{ton}$

j. Describe how the net present value, rate of return and economic price are used for economic decisions.

The net present value is used to compare among dependent alternatives. Projects are ranked by a company, and the net present values of the better projects are added to give a total that meets the capital available of investment.

The rate of return is used to compare independent alternatives. The rate of return for a project is compared with the rate of return from other investment alternatives and business opportunities such as bonds, stocks, real estate, purchase companies in other line of business.

The economic price is the price required to sell a product to make the projected rate of return. It is not related to the market price. If the market price is higher than the economic price, then a higher rate of return will be made.

23. Compute the net present value of these cash flows using a MARR of 15% and determine which schedule is better. Using Equation 14:

$$P = F(P/F, 15\%, n) = F(1 + i)^{-n} = F(1.15)^{-n}$$

n (year)	Case 1 F ($th/yr)	P/F $(1.15)^{-n}$	P ($th/yr)	Case 2 F ($th/yr)	P ($th/yr)
1	15	0.8696	13	15	13.0
2	50	0.7561	37.8	20	15.1
3	5	0.6575	3.3	25	16.4
5	10	0.5718	5.7	30	17.2
		Total	67.3		79.1

Case 1 is better than Case 2 and is preferred because the net present value of the cost is smaller.

24. 17. $0.5(\$6,500) + 0.10(\$6,000) + 0.15(\$6,250) + 0.25(6,700) = \$6,463$

25. Perform the following economic analysis and determine:

a. Net annual income before taxes = sales - total product cost = 21.4 - 19.3 = 2.1

b. Net annual cash flow before taxes = net annual income before taxes - annual capital expenditures for worn out equipment = 2.1 - 0.1 = 2.0

c. Depreciation = plant installed cost/economic life = 1.0/15 = 0.067

d. Taxes = tax rate*taxable income = tax rate*(net annual income before taxes- depreciation)

= 0.30*(2.1 - 0.067) = 0.61

e. Net annual income after taxes = net annual income before taxes – taxes

= 2.1 - 0.61 = 1.49

f. Net annual cash flow after taxes = net annual income after taxes - annual capital expenditures for worn out equipment

= 1.49 - 0.1 = 1.39

g. Net present value (NPV) based on the net annual cash flow after taxes and minim attractive rate of return.

$NPV = -CF_0 + A\{[(1 - (1+i)^{-n}]/i\} = - 4.0 + 1.39[1 - (1.15)^{-15}]/0.15 = 4.13$

h. Rate of return where the net present value is zero.

$NPV = -CF_0 + A\{[(1 - (1+i)^{-n}]/i\} = 0$

NPV(15%) = 4.13, NPV(30%) = 0.54, interpolating NPV(32.2%) = 0

i. Economic price. Economic price = (total product cost + annual cost of capital)/ product rate where annual cost of capital =EUAC = P*(A/P)

$$= P^*\{i/[1 - (1+i)^{-n}]\} = 4.0\{0.15/[1 - (1.15)^{-15}]\} = 0.68$$

Economic price = (19.3 + 0.1 + 0.68)/44.1 = $0.46 per pound

26. Economic Analysis for an Acetone Plant Design. Perform the following economic analysis and determine:

a. Net annual income before taxes = sales - total product cost = 28.0 - 26.5 = 1.5

b. Net annual cash flow before taxes = net annual income before taxes - annual capital expenditures for worn out equipment = 1.5 - 0.05 = 1.45

c. Depreciation = plant installed cost/economic life = 1.69/10 = 0.169

d. Taxes = tax rate*taxable income = tax rate*(net annual income before taxes- depreciation) = 0.35*(1.5 - 0.169) = 0.466

e. Net annual income after taxes = net annual income before taxes - taxes = 1.5 - 0.466 = 1.034

f. Net annual cash flow after taxes = net annual income after taxes - annual capital expenditures for worn out equipment = 1.034 - 0.05 = 0.984

g. Net present value (NPV) based on the net annual cash flow after taxes and minim attractive rate of return. NPV = $-CF_0 + A\{[(1 - (1+i)^{-n}]/i\}$

$$= - 2.6 + 0.984\{[1 - (1.15)^{-10}]/0.15\} = 2.34$$

h. Rate of return where the net present value is zero.

NPV = $-CF_0 + A\{[(1 - (1+i)^{-n}]/i\} = 0$ NPV(15%) = 2.41, NPV(30%) = 0.44
Interpolating NPV (34.7%) = 0

i. Economic price. Economic price =
(total product cost + annual equipment expenditures + annual cost of capital)/ product rate where annual cost of capital =EUAC = P*(A/P)

$$EUAC = P^*\{i/[1 - (1+i)^{-n}]\} = 2.61\{0.15/[1 - (1.15)^{-10}]\} = 0.52$$

Economic price = (26.5 + 0.05 + 0.52)/15.8 = $1.71 per kg

27. a. Equivalent uniform annual series, $EUAC = PW\{i/[(1 - (1+i)^{-n}]\}$ Equation. 20, where PW = $40,000, cost to regenerate the catalyst, negative; $i = 0.25$ and $n = 2$ years.

$$EUAC = \$40,000\{0.25/[1 - (1.25)^{-2}]\} = \$27,800$$

Net profit from regenerating catalyst = $135,000 - $27,800 = $107,200 in the year it is regenerated. Net profit from regenerating catalyst = $120,000 - $27,800 = $92,200 in the next year.

Year	1	2	3	4	5
Annual net profit	$135,000	$120,000/yr	$105,000/yr	$85,000/yr	$65,000

(no regeneration)

b. Annual profit for each year of the five years for year two

Regenerate in year 2	$135,000	$135,000 - $27,800 = $107,200	$120,000 - $27,800 $92,000	$105,000	$85,000

c Annual profit for each year of the five years for year three

Regenerate in year 3	$135,000	$120,000	$135,000 - $27,800 = $107,200	$120,000 - $27,800= $92,000	$85,000

d. Annual profit for each year of the five years for year four

Regenerate in year 4	$135,000	$120,000 $105,000	$135,000 - $27,800 = $107,200	$120,000 - $27,800= $92,000

e. Comparing annual net profit

Year	1	2	3	4	5	Total Profit
No regeneration	$135,000	$120,000	$105,000	$85,000	$65,000	$510,000
b Regenerate in 2	$135,000	$107,200	$92,200	$105,000	$85,000	$524,000
c Regenerate in 3	$135,000	$120,000	$107,200	$92,200	$105,000	$559,400
d Regenerate in 4	$135,000	$120,000	$105,000	$107,200	$92,200	$559,400

Regenerating in years 3 and 4 are the same. The best choice would be to invest in regeneration in year 4, delaying the expenditure.

28 See Turton, et al., 1998 for the detailed solution

Appendix A. Estimating Charts for Equipment, Plant and Manufacturing Costs

Cost estimates can be made rather quickly with estimating charts that are available from published sources and company past experience. Estimating charts are available from a number of sources including Hall, et.al, 1984, Douglas, 1988, Peters and Timmerhaus, 1991, and Garrett, 1989. The ones by Garrett are very complete, and he was careful to adjust the cost to 1987 with a chemical engineering index of 320. Several of the charts in Garrett have been reproduced in this appendix for major process equipment costs. See Figures A-1 through A-7. As Garrett points out that these charts give a "very rough cost estimate" for process equipment, and the price could "vary considerably" depending on exact design requirements, quality, maintenance and other factors. These charts are valuable for many preliminary design estimates, and engineers can use the average, representative equipment costs. Garrett states that actual prices can range from +100% to -50% depending on quality, designs and specifications, but the results obtained using these charts are in the generally correct range and represent good to high quality equipment. The alternatives to equipment estimation charts are to use capital cost estimation programs and manufactures' quotations, both of which are more expensive and time-consuming.

Along with the equipment estimating charts, Garrett has included charts for estimating complete plant costs such as the one reproduced here for C4 chemicals in Figure A- 8. For the plant cost estimating charts, this information was assembled from a number of sources, and Garrett states that the costs should be "roughly correct", and he recommends caution when using all of the charts. The costs include the plant and storage with a CE Index of 320. The alternative is to do a detailed plant cost estimate as listed in Table 2.

Along with the above estimating charts, Garrett has included charts for estimating manufacturing costs such as the one reproduced here for C4 and ammonia chemicals in Figure A- 9. For these estimating charts, this information was assembled from literature data, which he said was less that other cost data. Garrett states that the costs should be for "order of magnitude" estimates. In addition to the manufacturing cost versus plant capacity curves, Garrett includes tables for a few processes that has raw material and utility estimates which is useful for "conservative first approximations". A final table gives data on a single plant size with percentage breakdowns for raw material, depreciation, and utilities and labor, which can be used for a "rough estimate". The alternative is to do a total product cost estimate as listed in Table 3.

Estimating charts are log-log plots, and the cost data is approximately a straight line. These charts are based on the following scaling equation. The cost, C, can be predicted over a range of equipment size (capacity) values, S, if one value of C_1 and S_1 is known along with the exponent, n, Equation 1.

$$C = C_1 (S/S_1)^n \qquad (1)$$

where C_1 and S_1 are a known cost and equipment size (capacity). This equation has lead to a "0.6 rule" that n is approximately 0.6 as a rule of thumb. This exponent is given on some estimating charts. In Perry's, 1997 Table 25-49, an extensive listing of equipment is given with sizes, costs and exponents, so the above equation can be used to estimate purchased cost. This equation is used to include the effects on temperature, pressure, materials of construction (Table

4), inflation (Figure 8 and 9), and location in the world (Figure 10) as described below. McConnell, 2009, gives additional data for international locations based on 2009 data.

Some of the charts include the instillation factor that is a multiplier for the equipment cost to estimate the installed cost of the equipment. Some charts may include the module factor that is a multiplier of the equipment cost to give the purchase and instillation of major equipment along with supporting equipment. Some charts may include a factor to estimate the cost of, for example, stainless steel. For the plant cost estimating charts, this information was assembled from a number of sources, and Garrett states that the costs should be "roughly correct".

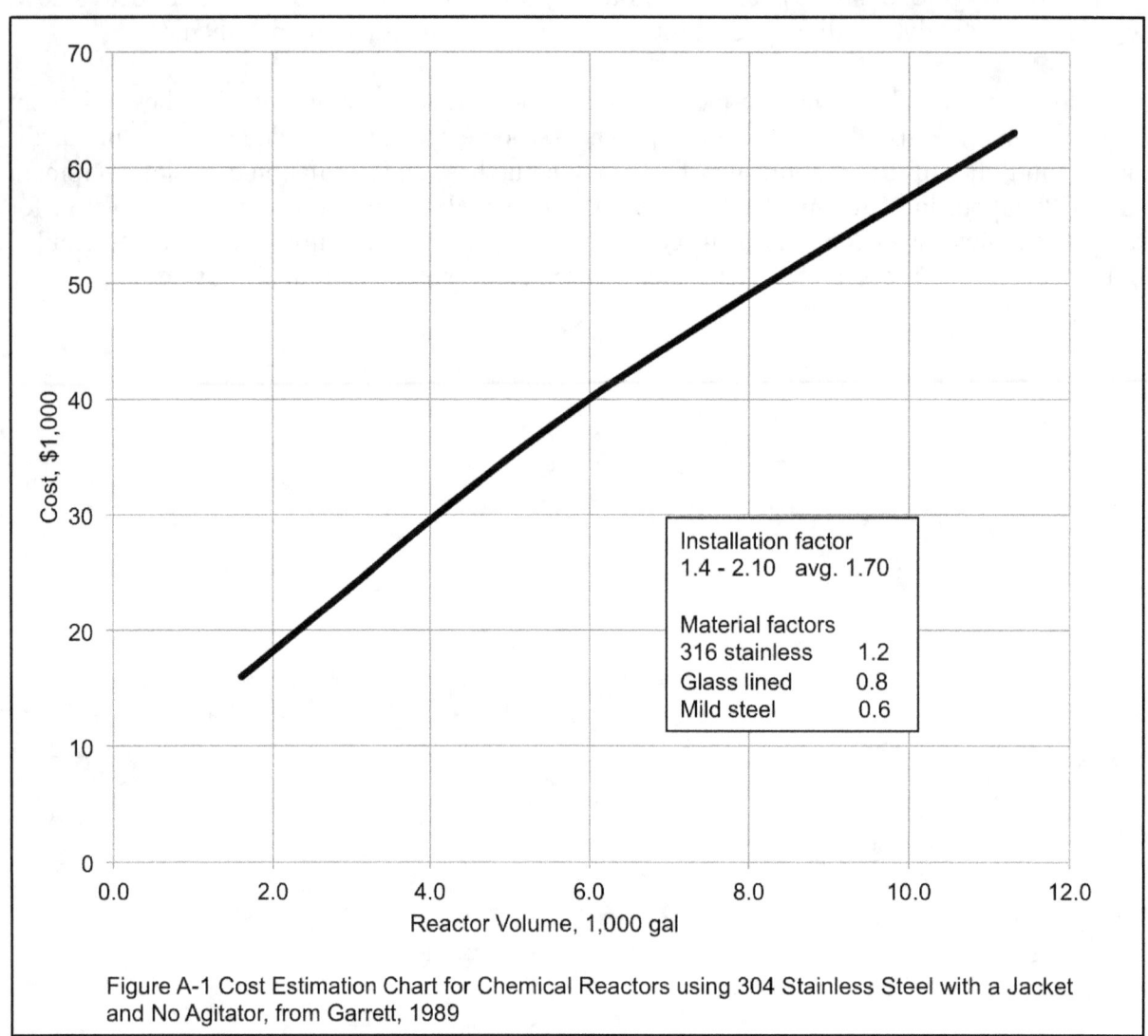

Figure A-1 Cost Estimation Chart for Chemical Reactors using 304 Stainless Steel with a Jacket and No Agitator, from Garrett, 1989

150

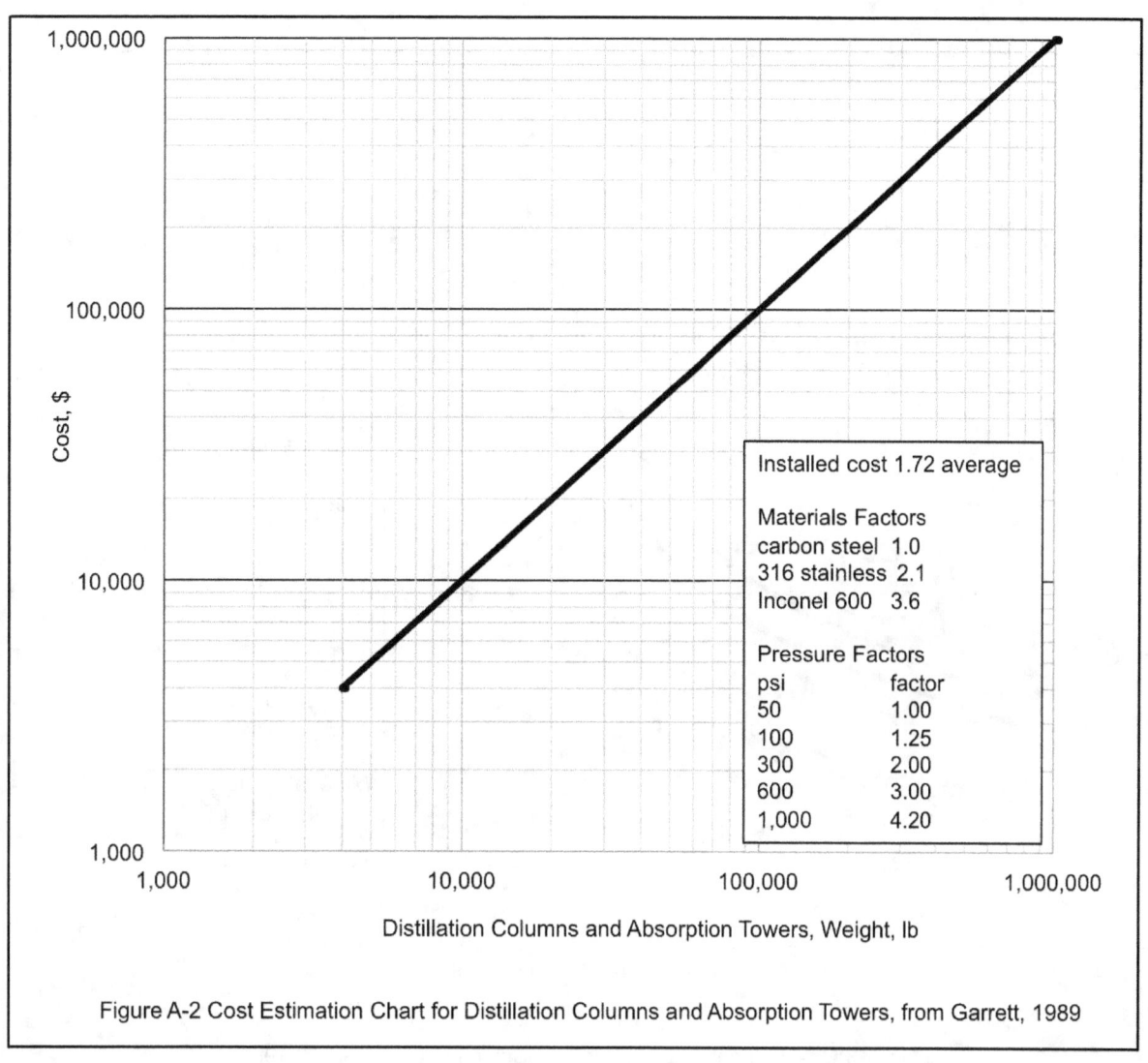

Figure A-2 Cost Estimation Chart for Distillation Columns and Absorption Towers, from Garrett, 1989

151

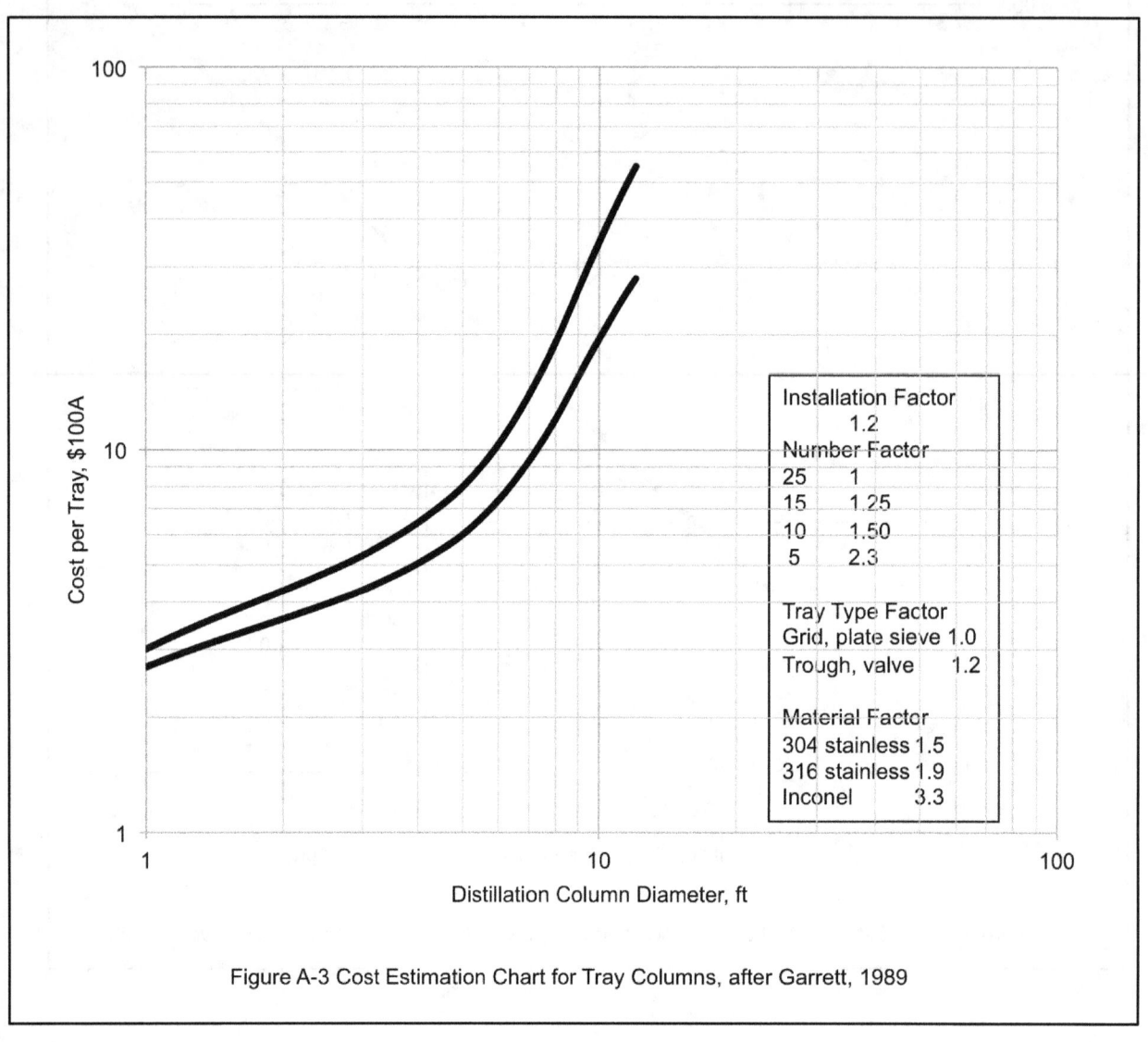

Figure A-3 Cost Estimation Chart for Tray Columns, after Garrett, 1989

152

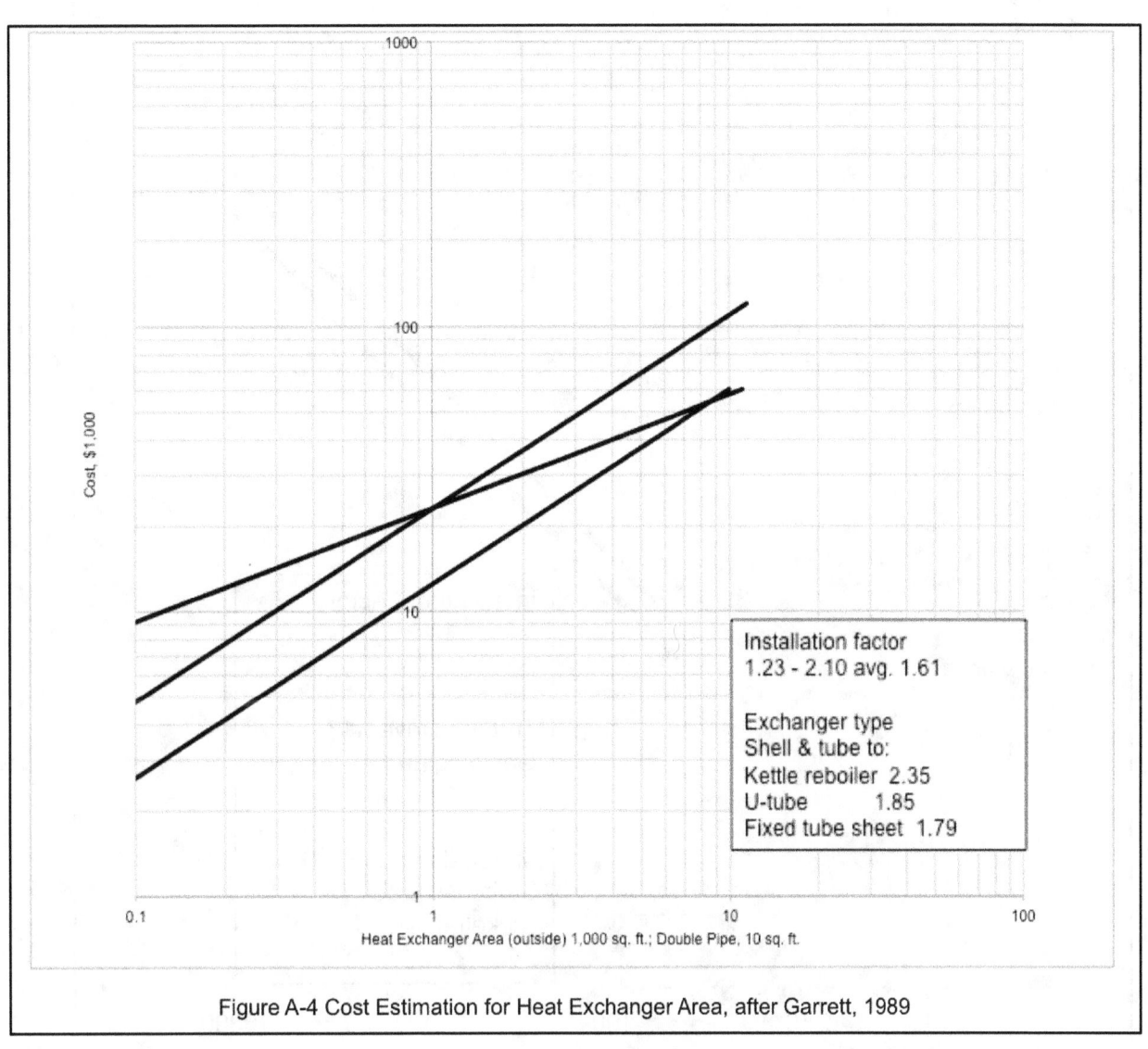

Figure A-4 Cost Estimation for Heat Exchanger Area, after Garrett, 1989

153

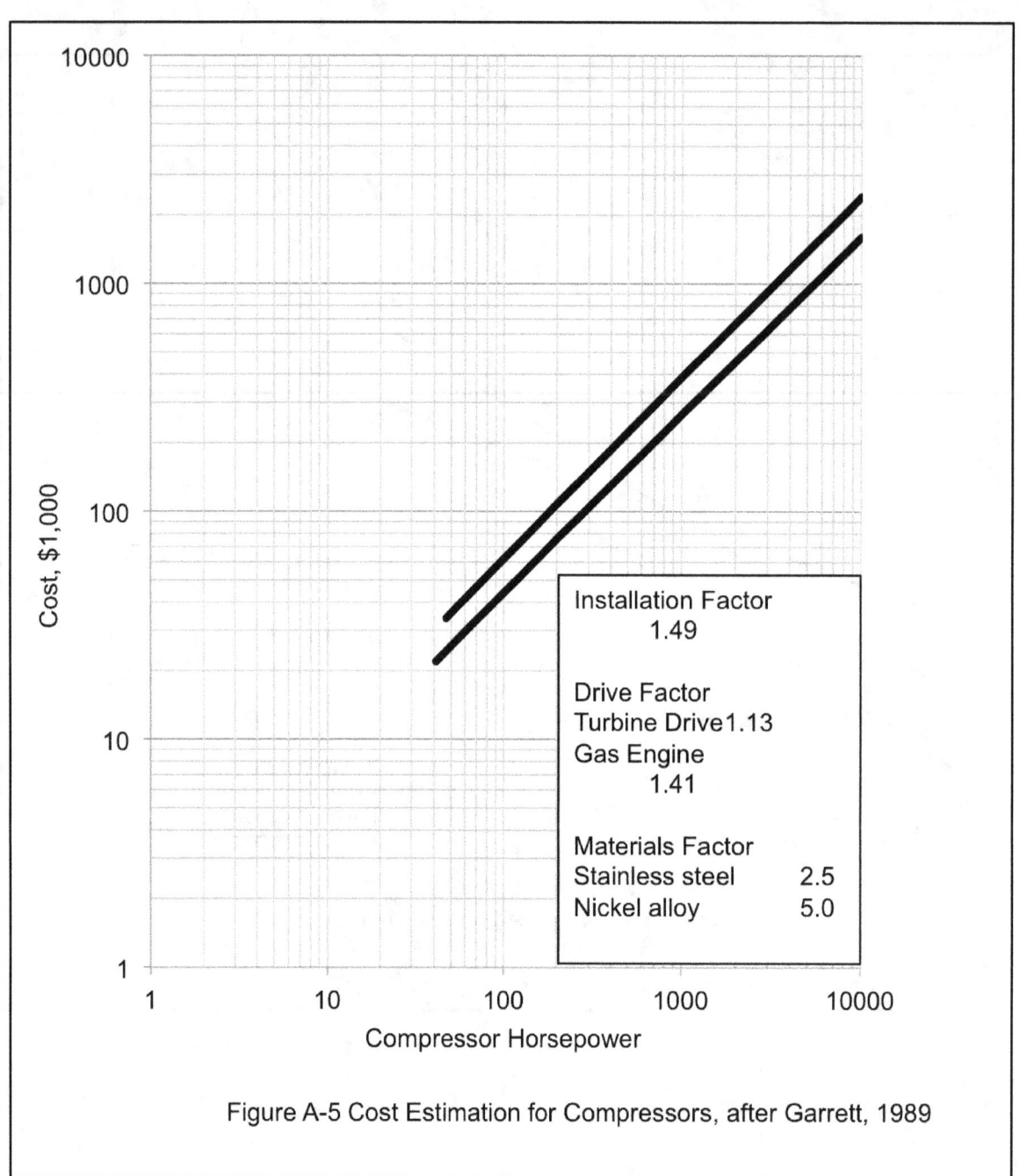

Figure A-5 Cost Estimation for Compressors, after Garrett, 1989

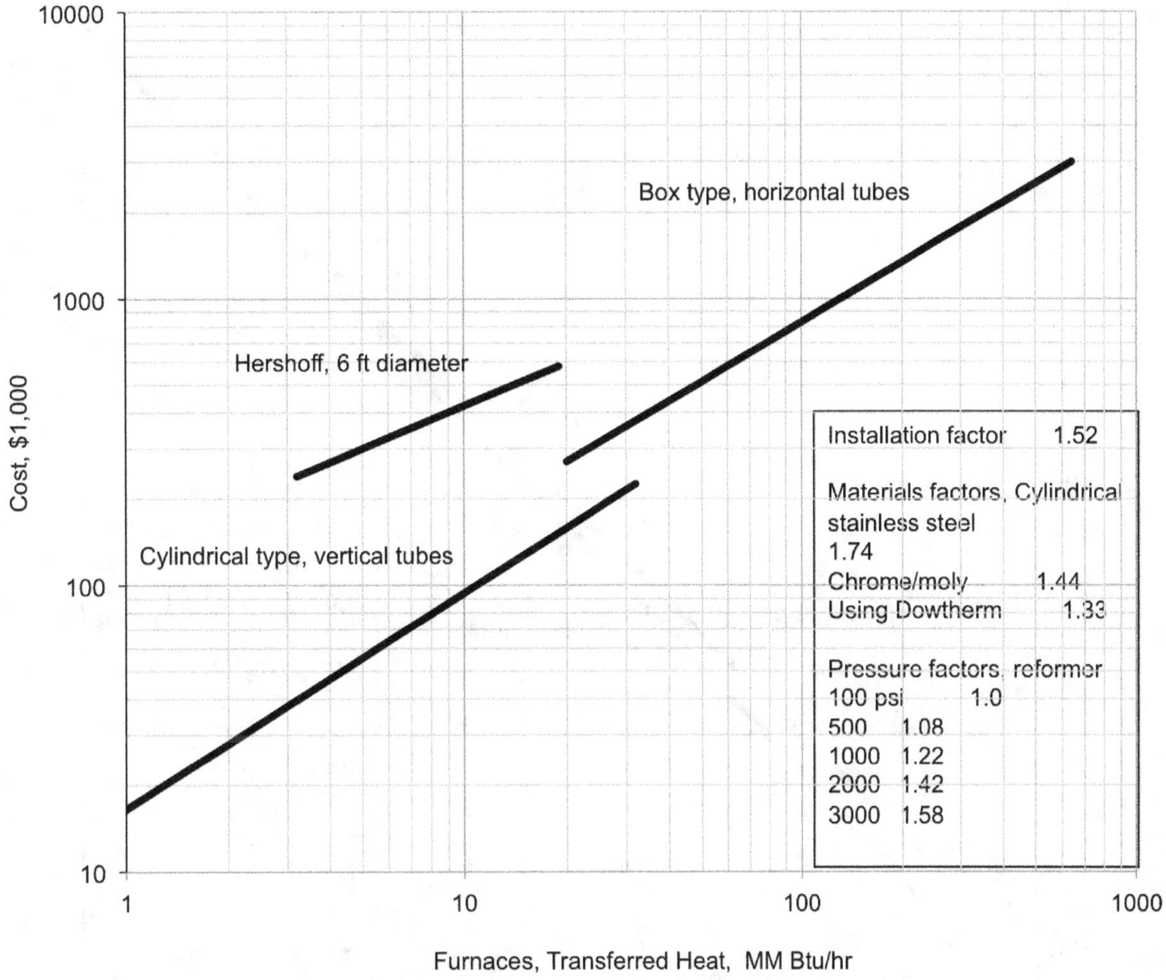
10000
1000
100
10
Cost, $1,000
1
10
100
1000
Furnaces, Transferred Heat, MM Btu/hr
Box type, horizontal tubes
Hershoff, 6 ft diameter
Cylindrical type, vertical tubes
Installation factor 1.52
Materials factors, Cylindrical
stainless steel
1.74
Chrome/moly 1.44
Using Dowtherm 1.33
Pressure factors, reformer
100 psi 1.0
500 1.08
1000 1.22
2000 1.42
3000 1.58

Figure A-6 Cost Estimation for Furnaces, after Garrett, 1989

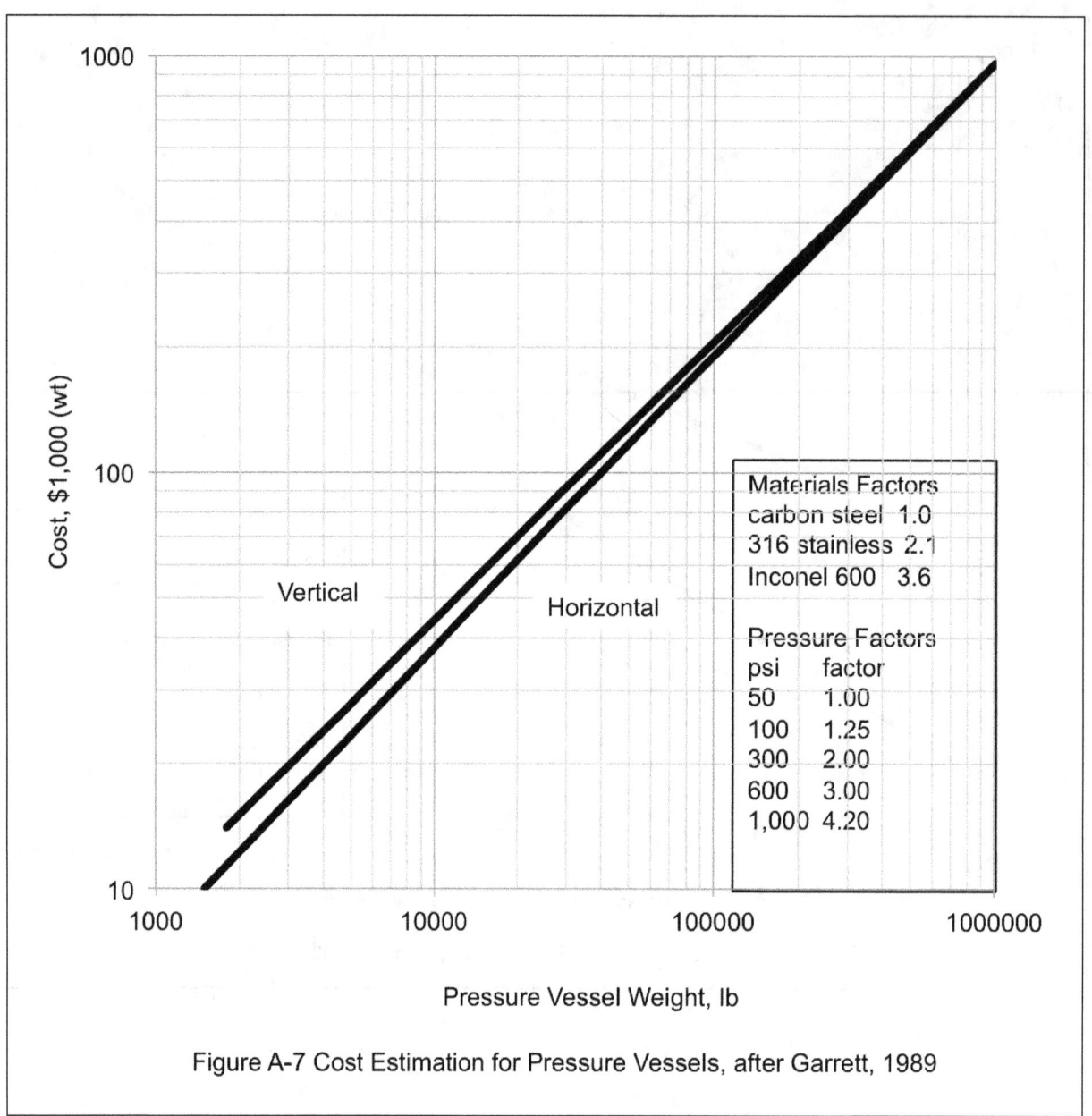

Figure A-7 Cost Estimation for Pressure Vessels, after Garrett, 1989

156

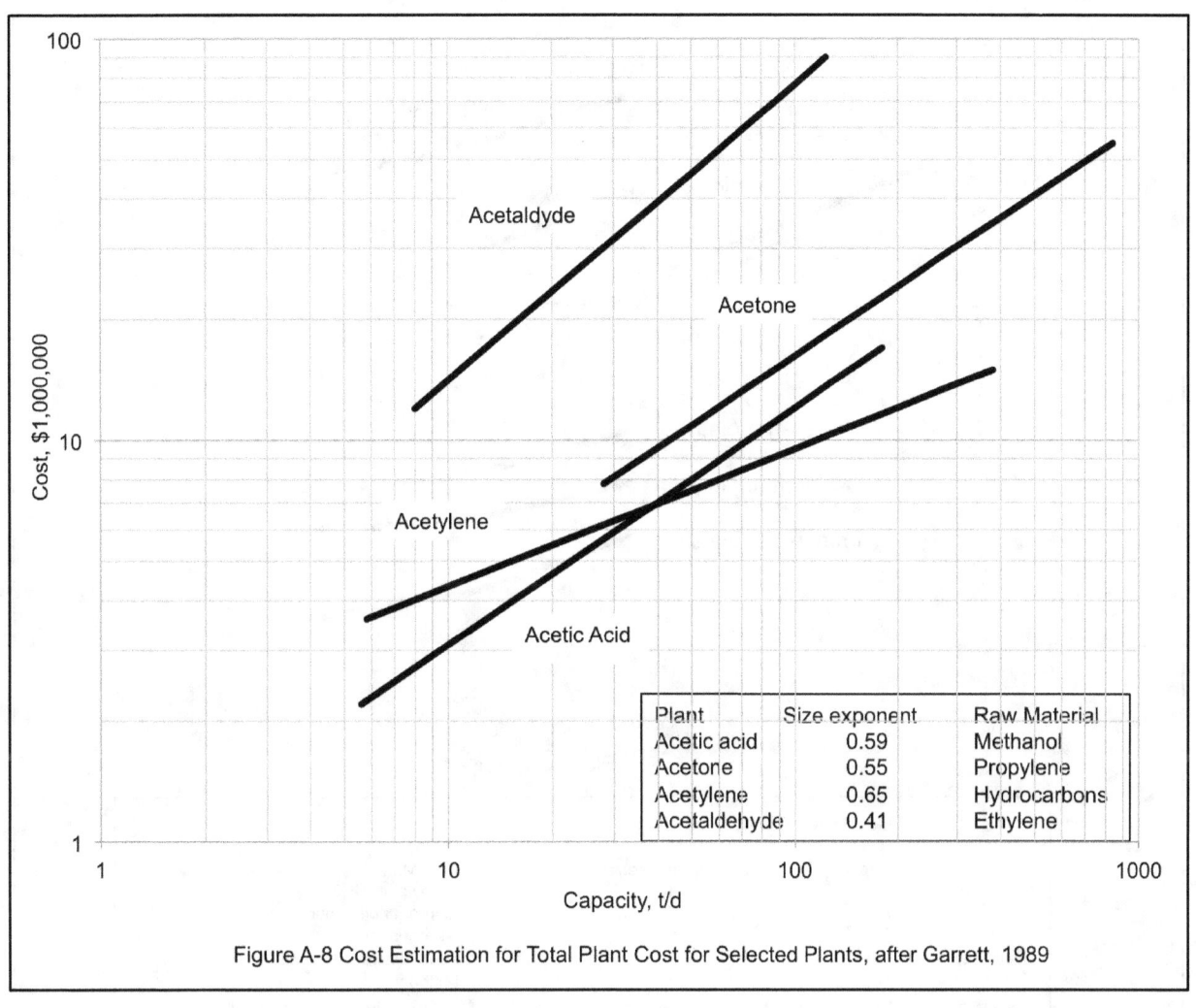

Plant	Size exponent	Raw Material
Acetic acid	0.59	Methanol
Acetone	0.55	Propylene
Acetylene	0.65	Hydrocarbons
Acetaldehyde	0.41	Ethylene

Figure A-8 Cost Estimation for Total Plant Cost for Selected Plants, after Garrett, 1989

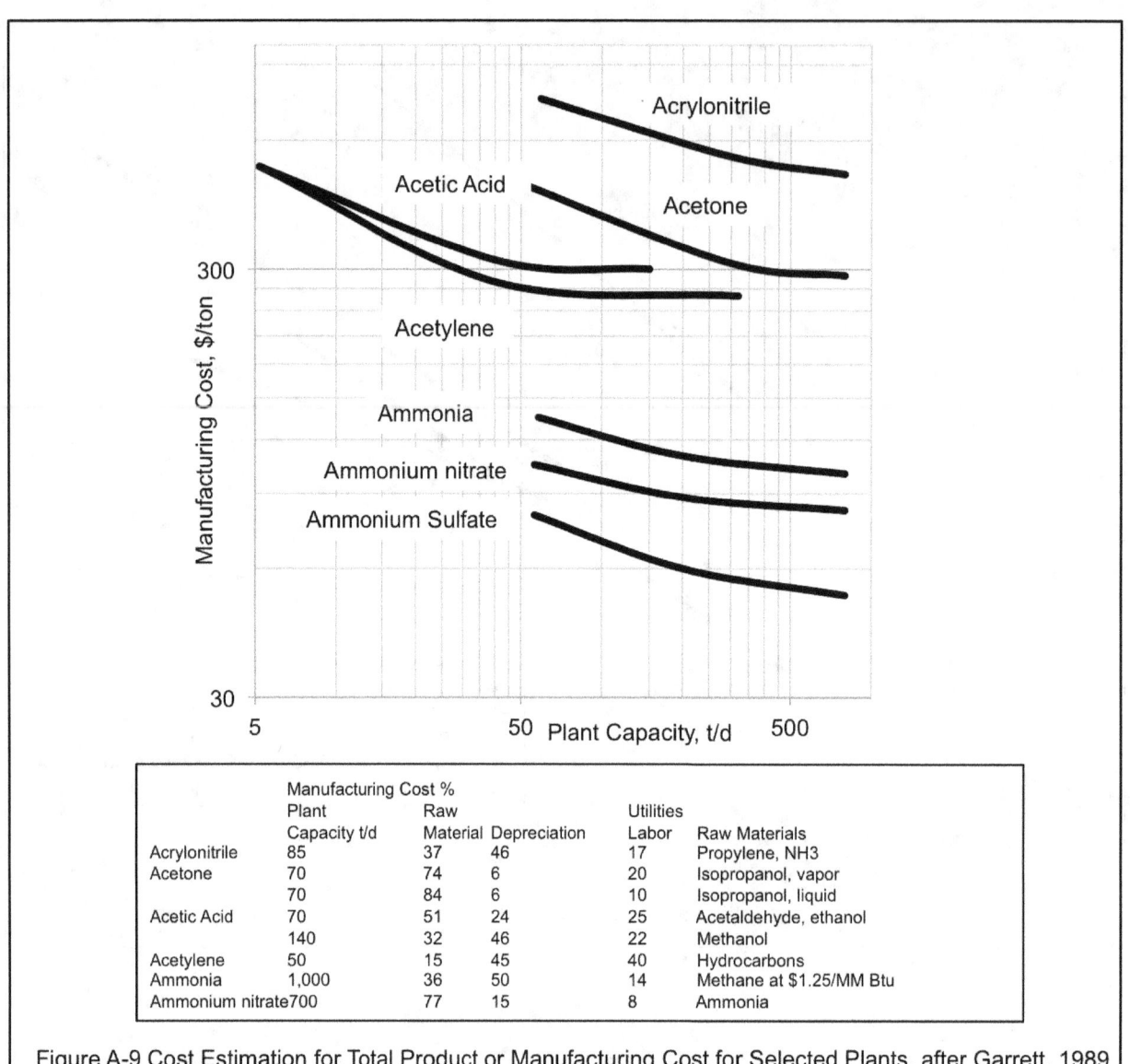

Figure A-9 Cost Estimation for Total Product or Manufacturing Cost for Selected Plants, after Garrett, 1989

Appendix B. Supply, Demand and Price Elasticity (after Sengupta and Pike, 2012)

Introduction

Describing supply and demand of goods and services in the economy is complicated. To provide insight, two measured parameters are used: price elasticity of demand and price elasticity of supply. In economics, elasticity is defined as the ratio of the percent change in one variable to the percent change in another variable.

Price elasticity of demand (PED) is a measure of the responsiveness of quantity demanded to changes in price (Arnold, 2008) as shown in Equation B-1. Mathematically, it is the ratio of percent change in a quantity of goods or services, Q, to the percent change in price, P. It shows the response of a quantity demanded for goods or services to a change in price. PED is almost always negative, i.e. an increase in price will cause a reduction in demand. A value of PED between zero and minus one is considered inelastic.

$$PED = \frac{\Delta Q / Q \bullet 100}{\Delta P / P \bullet 100} \qquad (B\text{-}1)$$

Economists often use the absolute value of price elasticity for analysis. The value of PED can be interpreted as follows.

- If PED < -1 then Demand is Price Elastic (Quantity demanded changes proportionately more than price changes)
- If PED = -1 then Demand is Unit Elastic (Quantity demanded changes proportionately to price changes)
- If PED >-1 then Demand is Price Inelastic (Quantity demanded changes proportionately less than price changes)
- If PED = 0 then Demand is Perfectly Inelastic (Quantity demanded does not change as price changes).
- If PED = - ∞ then Demand is Perfectly Elastic (Quantity demanded is extremely responsive to even small changes in price).

Price elasticity of supply (PES) is the ratio of percent change in a quantity supplied, S, to the percent change in price, P as shown in Equation B-2. It measures the sensitivity of the quantity of goods and services to the change in market price for those goods or services. PES is almost always positive, i.e. an increase in price will cause an increase in supply. A value of PES less than one is considered inelastic.

$$PES = \frac{\Delta S / S \bullet 100}{\Delta P / P \bullet 100} \qquad (B\text{-}2)$$

The value of PES can be interpreted as follows.
- If PES > 1 then Supply is Price Elastic (Supply quantity changes proportionately more than price changes)

- If PES = 1 then Supply is Unit Elastic (Supply quantity changes proportionately to price changes)
- If PES < 1 then Supply is Price Inelastic (Supply quantity changes proportionately less than price changes)
- If PES = 0 then Supply is Perfectly Inelastic (Supply quantity does not change as price changes).
- If PES = ∞ then Supply is Perfectly Elastic (Supply quantity is extremely responsive to even small changes in price).

Other parameters measured to describe economic interactions are given below. They have definitions similar to PED and PES. More information is provided in Elasticity (economics) Wikipedia, 2010. The Table B-1 gives some values of price elasticity of demand and price elasticity of supply reported in Elasticity (economics) Wikipedia, 2010(c).

- Income elasticity of demand
- Cross price elasticity of demand
- Cross elasticity of demand between firms
- Elasticity of intertemporal substitution
- Elasticity of scale

Table B-1 Some Typical Values of Price Elasticity of Demand and Price Elasticity of Supply (Wikipedia, 2010)

Price Elasticity of Demand		Price Elasticity of Supply	
Oil (world)	-0.4	Heating oil	1.57
Gasoline	-0.25 to -0.64	Gasoline	1.61
Transportation	-0.20 (bus) to -2.8 (car)	Housing	1.6 to 3.7
Steel	-0.2 to -0.3	Steel	1.2
Rice	-0.47	Cotton	0.3 to 1.0
Livestock	-0.5 to -0.6	Tobacco	7.0
Airline travel	-0.3 to -1.5		

The method for determining price elasticity of supply and demand, as defined above, can be applied to raw materials, products and intermediates in chemicals manufacturing. The market for fermentation ethanol (also known as bioethanol) as a chemical is not yet established in the United States. However, the potential for bioethanol as a chemical exists. The price elasticity of demand and supply of corn (raw material), the price elasticity of demand and supply for bio-ethanol (intermediate for future petroleum ethanol substitute), and the price elasticity of demand for ethylene (current petroleum feedstock based ethylene) are estimated in this section, so that an insight can be gained for the requirements to evaluate these parameters.

Price Elasticity of Supply and Demand for Corn

Historically, corn has been used for food use and feed grain. With the ongoing efforts to substitute fossil fuels with biofuels, there has been a rise in the importance of fuel use of cereals (Banerjee, 2010). To study the effect of the rise in demand of corn, price elasticity was used. The data for price, supply and demand for corn was obtained from USDA, 2010 (Figures B-1 and B-2.

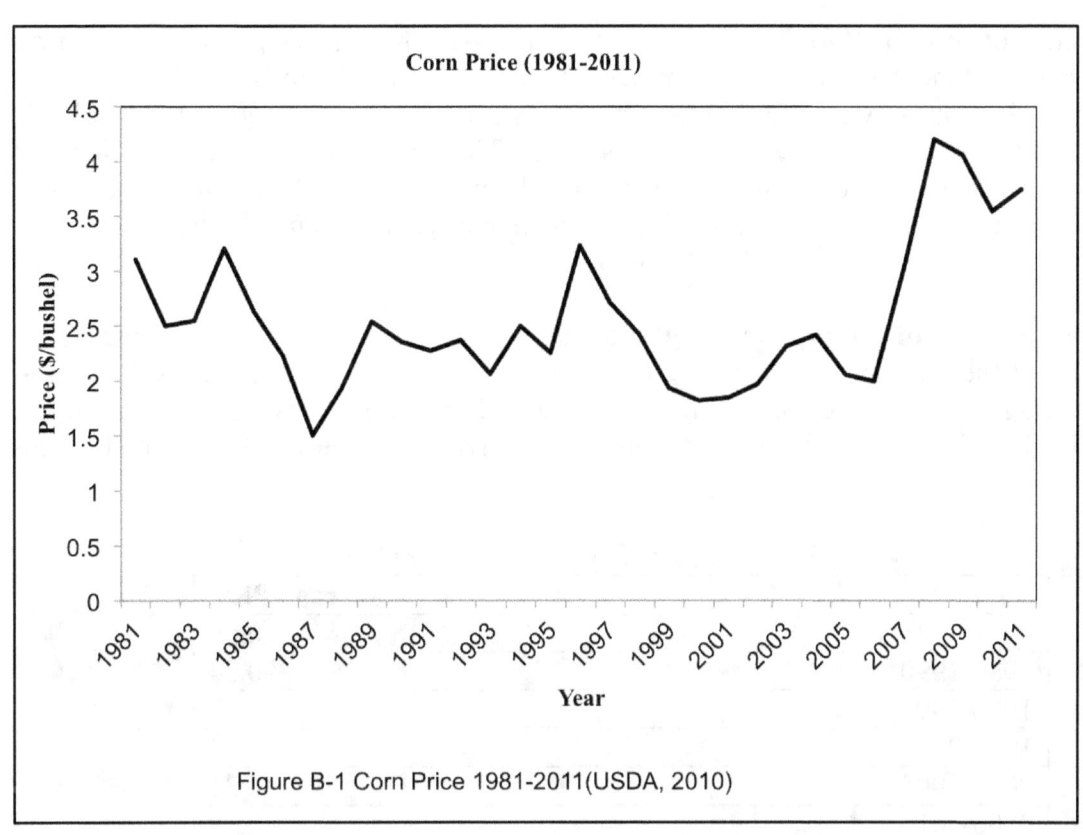

Figure B-1 Corn Price 1981-2011(USDA, 2010)

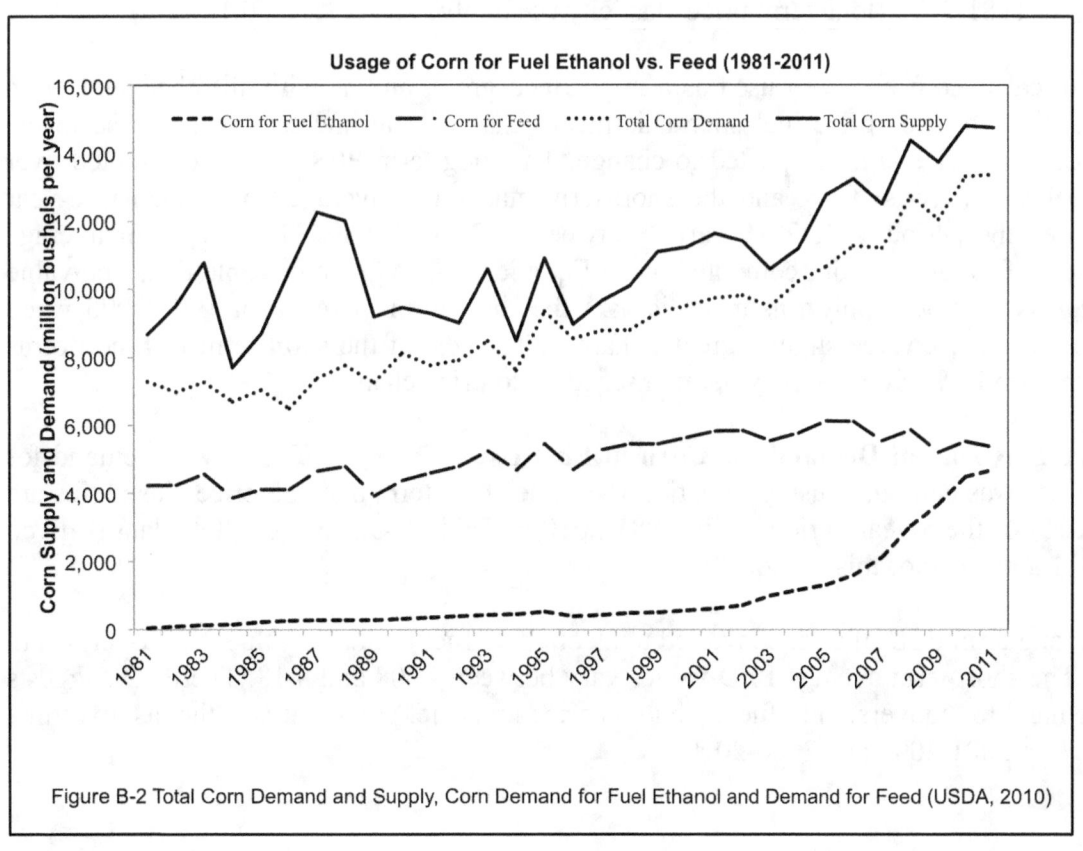

Figure B-2 Total Corn Demand and Supply, Corn Demand for Fuel Ethanol and Demand for Feed (USDA, 2010)

The price of corn for 1981-2011 is given in Figure B-1. The total supply of corn, the total demand for corn and the demand for fuel and feed use are given in Figure B-2. The short term and long term price elasticity for the supply of corn was computed for the period 1981-2011, in intervals of five years for the short term price elasticities. The price elasticity of demand for corn in alcohol fuel use and for feed use was computed for the period 1981-2011, in intervals of five years for the short-term price elasticities. Also, for price elasticity of demand, the annual price elasticity data is given for the period of 2000-2010.

Price Elasticity of Supply for Corn: The supply for corn was distributed primarily for the production of fuel ethanol and the production of feed. Other uses of corn include food uses, alcoholic beverages and seed uses. The long term and short term average price elasticity of supply and is given in Table B-2. The complete data is given in Table B-10 at the end of this section.

Table B-2 Price Elasticity of Supply for Corn

Year	Average PES
1981-1985	0.59
1986-1990	-0.24
1990-1995	-1.29
1996-2000	0.12
2001-2005	-0.31
2006-2010	0.18
1981-2010 (long term price elasticity of supply)	-0.16

Historically, corn has been used as a major feed grain, and as with all other feed grain crops, it is expected to be price inelastic. This means that with an increase or decrease in the supply of corn, the price is not expected to change. The long term PES of corn calculated over the period 1981-2010 was -0.16, and the short-term elasticities averaged over each five-year period between the period 1981-2010 varied between -1.29 and 0.59. The results for average price elasticity of supply for corn computed using Equation B-2 is given in Table B-2. The value of PES suggests that the supply was inelastic over the short terms, except for 1990-1995 when the supply was -1.29, a value slightly greater than -1. The rest of the short term PES conforms with the long term PES of corn supply being insensitive to price changes.

Price Elasticity of Demand for Corn in Fuel Use: The price elasticity of demand for corn in fuel use was computed using Equation B-1. The short-term average price elasticities are shown for each of the 5-year periods from 1981-2010 in Table B-3. The complete data is given in Table B-11 at the end of this section.

The long term PED of corn converted to alcohol for fuel use has a value 0.51 as shown in Table D-3. The short-term average PED values vary between -1.08 and 4.18. The PED suggests that corn demand for conversion to fuel alcohol is price inelastic, particularly in the last two time periods between 2001-2005 and 2006-2010.

162

Table B-3Average Price Elasticity of Demand for Corn Used for Alcohol

Corn Used for Fuel Alcohol Production	
Year	Average PED
1981-1985	4.18
1986-1990	-0.89
1990-1995	0.22
1996-2000	0.64
2001-2005	0.01
2006-2010	-1.08
1981-2010 (long term price elasticity of demand)	0.51

To study the PED in these two time periods, the demand, price and PED for each year are shown in Table B-4. The annual price elasticities reveal that the PED varied from -7.27 to 4.01 in the year range 2001-2005 and -6.18 to 1.15 in the year range 2006-2010. These high values in PED shows that the demand in corn uses for alcohol production have increased steadily. Also, the change in demand was never negative for the corn use in fuel. The price for corn, has fluctuated in this time, varying from $1.85/bushel in 2001 increasing to $2.42/bushel in 2004 and decreasing to $2.00/bushel in 2006 and again increasing to $4.20/bushel in 2008.These results show that the demand of corn for alcohol use has been price elastic over the last ten years.

Table B-4 Annual Price Elasticity of Demand for Corn Used for Alcohol from 2001-2010

Year	Production (million bushels per year)	ΔQ/Q*100	Price ($/bushel)	ΔP/P*100	PED
2001	629.83	12.29	1.85	6.49	1.89
2002	707.24	40.76	1.97	17.77	2.29
2003	995.50	17.28	2.32	4.31	4.01
2004	1,167.55	13.33	2.42	-14.88	-0.90
2005	1,323.21	21.17	2.06	-2.91	-7.27
				Average PED	0.01
2006	1,603.32	32.19	2.00	52.00	0.62
2007	2,119.49	43.87	3.04	38.16	1.15
2008	3,049.21	20.58	4.20	-3.33	-6.18
2009	3,676.88	22.39	4.06	-12.56	-1.78
2010	4,500.00	4.44	3.55	5.63	0.79
				Average PED	-1.08

Price Elasticity of Demand for Corn in Feed Use: The price elasticity of demand for corn in feed use was computed using Equation B-1. The short-term average price elasticities are shown for each of the 5-year periods from 1981-2010 in Table B-5. The complete data is given in Table B-12 at the end of this section.

Table B-5 Average Price Elasticity of Demand for Corn Used for Feed

Corn Used for Feed	
Year	Average PED
1981-1985	0.59
1986-1990	-0.80
1990-1995	-0.46
1996-2000	0.06
2001-2005	0.09
2006-2010	0.45
1981-2010 (long term price elasticity of demand)	-0.01

The long term PED of corn used as feed has a value -0.01 as seen from Table B-5. This confirms the general notion that the price of feed-grains is perfectly inelastic, meaning that the demand for feed is not going to decrease even for changes in price. This can happen due to several factors, one of them being the unavailability of alternatives (Banerjee, 2010). From Table B-5, it is seen that the value of PED ranges from -0.80-0.59.

The result for annual price elasticity for the ten-year period from 2001-2010 is given Table B-6. This table also shows that the corn used for feed is price inelastic for each year except for the year 2008. A closer look at the quantity demanded in the year 2009 shows that the quantity demanded for feed dropped in 2009 for a 38% increase in price from 2007 to 2008 for corn. This is the only instance in the time period when the price of corn reached a record high of $4.20/bushel of corn, and that reflected in the PED for corn used as feed.

Table B-6 Annual Price Elasticity of Demand for Corn Used for Feed from 2001-2010

Year	Production (million bushels per year)	$\Delta Q/Q*100$	Price ($/bushel)	$\Delta P/P*100$	PED
2001	5822.05	0.46	1.85	6.49	0.07
2002	5848.75	-5.14	1.97	17.77	-0.29
2003	5548.31	4.20	2.32	4.31	0.97
2004	5781.24	6.12	2.42	-14.88	-0.41
2005	6135.08	-0.33	2.06	-2.91	0.11
				Average PED	0.09
2006	6115.06	-9.40	2.00	52.00	-0.18
2007	5540.13	5.73	3.04	38.16	0.15
2008	5857.74	-11.14	4.20	-3.33	3.34
2009	5205.28	6.14	4.06	-12.56	-0.49
2010	5525.00	-3.17	3.55	5.63	-0.56
				Average PED	0.45

An insight into this period reveals that the demand for corn used for alcohol production was gaining impetus during this period, and the demand for feed production remained fairly constant. The cost for corn remained low, enabling a higher market for alcohol production from corn. The results from the price elasticity analysis suggests that corn production and demand for ethanol is highly elastic to changes in corn prices, whereas the market for feed is generally inelastic to price changes in corn (Banerjee, 2010).

Price Elasticity of Supply and Demand for Ethanol

Supply and demand elasticities in the U. S. ethanol fuel market have been evaluated by Luchansky and Monks, 2009 with data from various sources for the periods shown in Table B-7 below. For their demand model, ethanol demand elasticity was -1.605 to -2.915 during the period Jan 1984 to Dec 1987 and -0.417 to -1.503 for Jan. 1998 to May 1993. They offered the explanation of very price elastic for demand being caused by the changing availability of gasoline additives such as MTBE. Based on their supply model, ethanol price elasticity of supply ranged from 0.224 to 0.258 during the period Jan 1984 to Dec 1987 and was 0.044 for Jan. 1988 to May 1993, essentially inelastic. They reported that ethanol production was running largely at capacity during these periods. Effects of corn and gasoline supply and demand were discussed in relation to these price elasticities.

Table B-7 Ethanol Supply and Demand Price Elasticity (Luchansky and Monks, 2008)

	Price Elasticity of Demand	Price Elasticity of Supply
Jan, 1984 to Dec. 1987	-1.605 to -2.915	0.224 to 0.258
Jan. 1988 to May 1993	-0.417 to-1.503	0.043 to 0.044

Ethanol elasticity of supply, PES, was estimated for 2009 using data from the Commodities Report of *Ethanol Producers Magazine*, (Kment, 2009), and the results are given in Table B-8. The average was 0.425 with all of the values less than 1.71, implying inelasticity. During this period the Commodities Report describe the market in terms like: "gasoline and ethanol markets continue to gain light support," current production level of ethanol is enough to handle current and expected demands," "prices bounced 20 to 30 cents higher from mid-April to mid-May and have the potential to increase an additional 20 to 30 cents throughout the summer," "ethanol prices have weakened significantly through the first half of the summer," "overall demand is expected to remain stable to strong over the near future," "over the past several months, corn prices have had a great impact on the price of ethanol, giving the more of a 'cost-plus' feel than a true companion market to gasoline," In Figure B.3 the production of ethanol is shown from November, 2007 to October, 2009, and in Figure B.4 the price of ethanol is shown from March, 2009 to January, 2010 from *Ethanol Producers Magazine*, (Kment, 2009).

Table B-7 Estimation of the Elasticity of Supply for Ethanol in 2009 (Kment, 2009)

Date	Production (bbl/day)	ΔQ/Q	Price ($/gal)	ΔP/P	PES
Mar-09	669,000	0.0000	1.67	-0.012	0.000
Apr-09	669,000	0.0000	1.65	0.055	0.000
May-09	669,000	0.0374	1.74	0.034	1.084
Jun-09	694,000	0.0476	1.8	0.028	1.712
Jul-09	727,000	-0.0028	1.85	-0.027	0.102
Aug-09	725,000	0.0000	1.8	-0.028	0.000
Sep-09	725,000	0.0221	1.75	0.286	0.077
				Avg PES	0.231

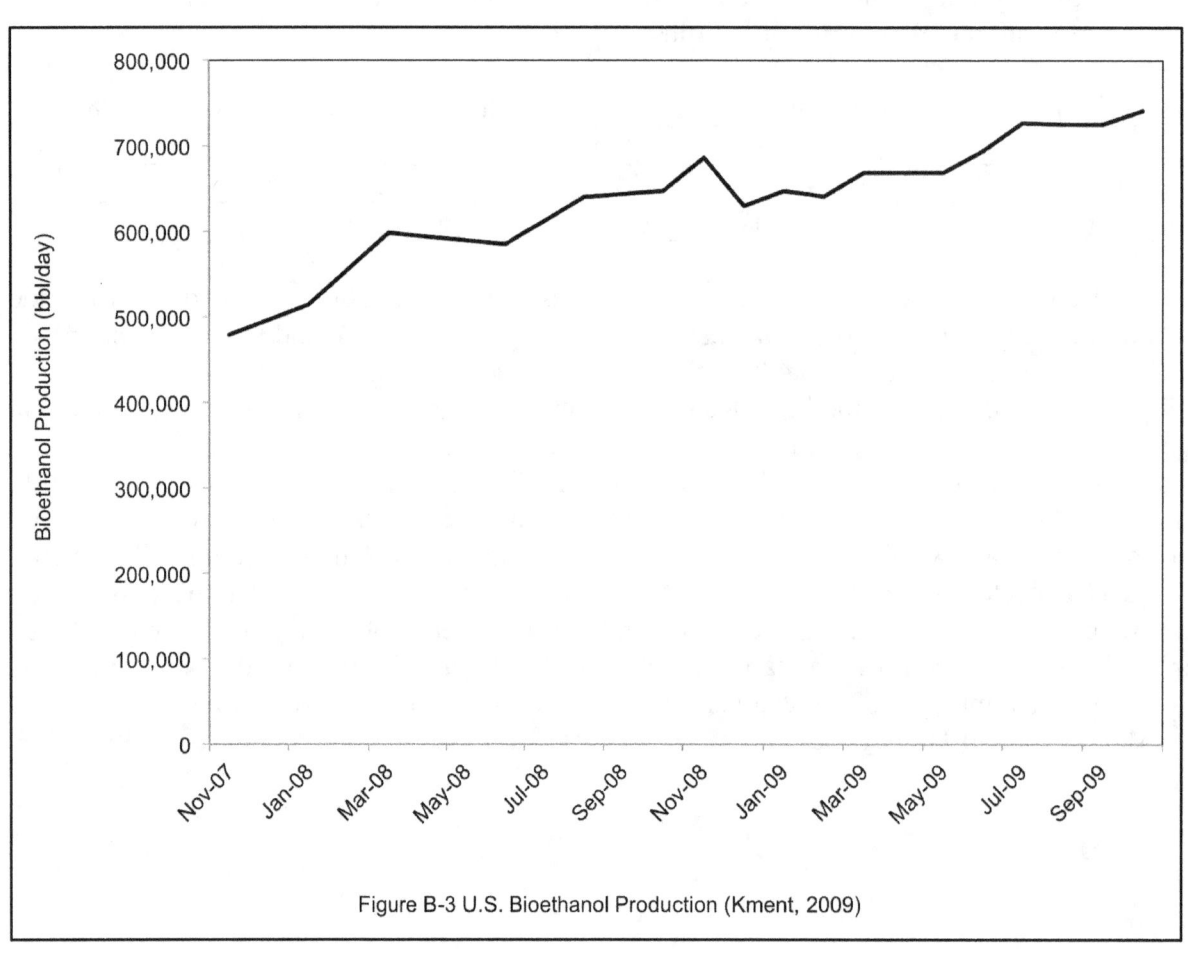

Figure B-3 U.S. Bioethanol Production (Kment, 2009)

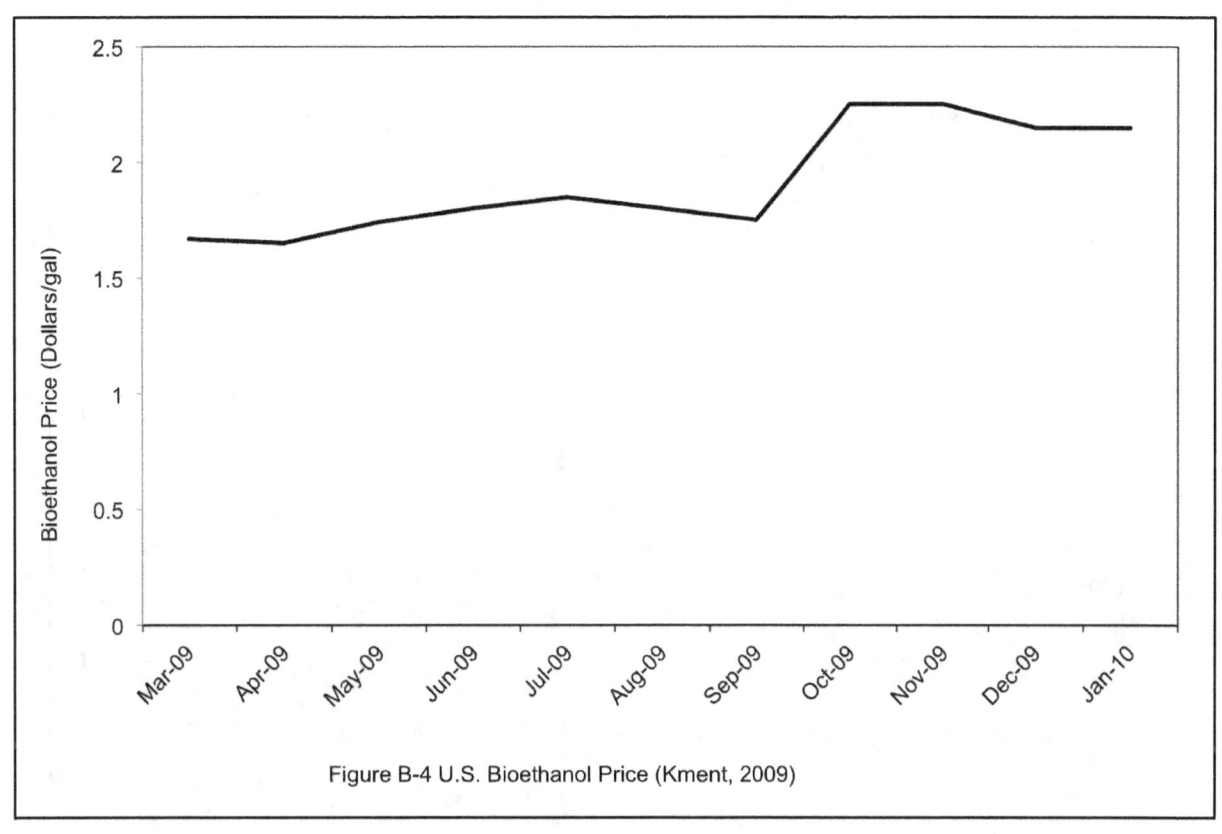

Figure B-4 U.S. Bioethanol Price (Kment, 2009)

Price Elasticity of Demand for Ethylene

The price elasticity of demand for ethylene was estimated using limited data available from C&E News, 2009. See Table 8. Ethylene production is shown in Figure B-5 and ICIS, 2008. Ethylene prices in the United States are shown in Figure B-6, as shown in Table B-9. The value of -0.416 for the price elasticity of demand was the result of a decrease in price that resulted in an increase in demand. The ICIS Chemical Business, 2009 reported that during this period "buyers pushed for a decrease in price on the heels of ample supply and lower production costs."

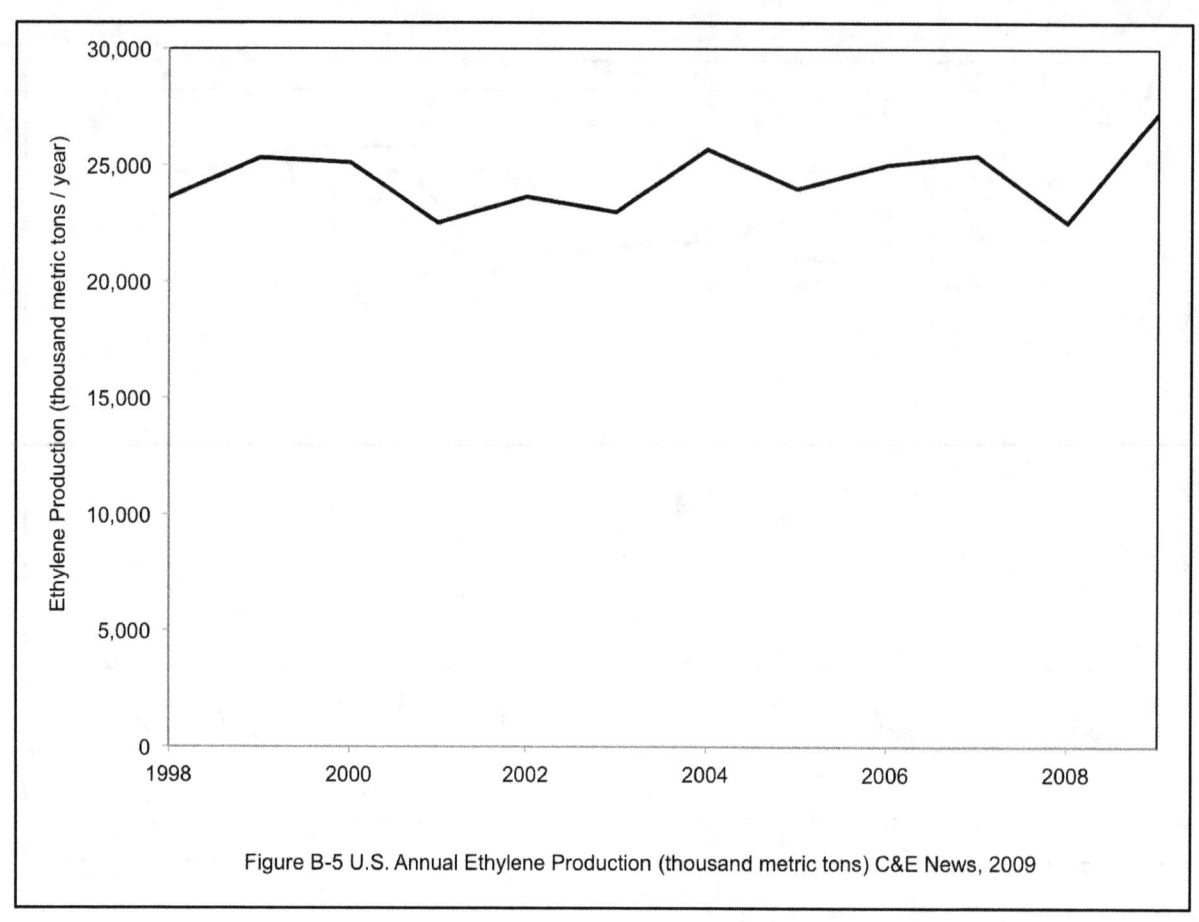

Figure B-5 U.S. Annual Ethylene Production (thousand metric tons) C&E News, 2009

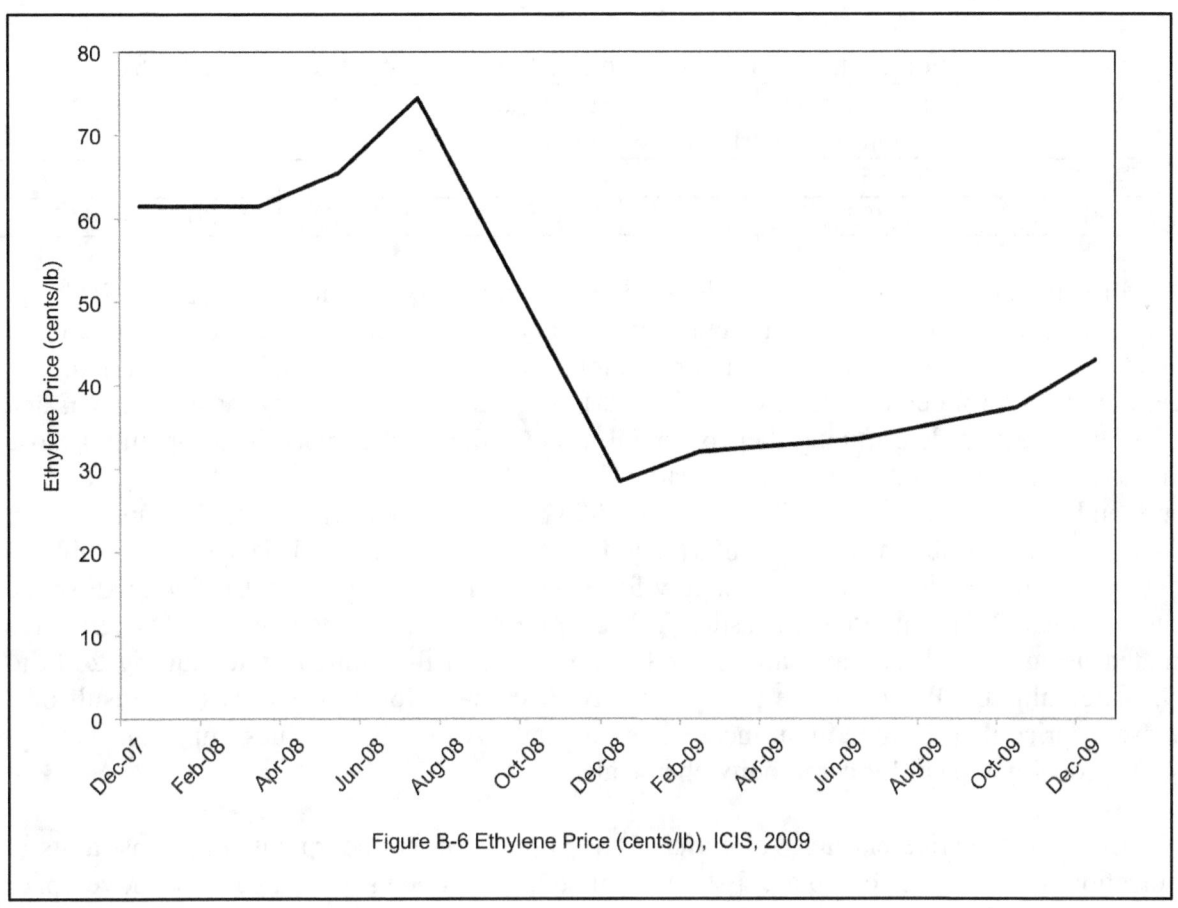

Figure B-6 Ethylene Price (cents/lb), ICIS, 2009

Table B-8 Estimation of the Elasticity of Demand for Ethylene for the period 2008-2009

Year	Ethylene Production (thousand metric tons)	Ethylene Price (cents/lb)	$\Delta P/P$	$\Delta Q/Q$	PED
2008	22,554	65.75			
2009	27,252	32.81	0.2083	-0.500	-0.416

In summary, price elasticity of demand (PED) and price elasticity of supply (PES) are useful measures of economic activity, but there is very limited data available to evaluate these parameters. Having values for chemicals from biomass, such as ethanol and ethanol derivatives, glycerol and glycerol derivatives, acetic acid and fatty acid methyl and ethyl esters (FAME and FAEE), would aid in determining their potential to enter the market place in competition those from petroleum. The only detailed evaluation by Luchansky and Monks, 2009 was for bio-ethanol and it was for the period from 1988 to 1994 where ethanol price elasticity ranged from 0.224 to 0.258. To obtain more recent values for ethanol, very limited data was available in 2009 to evaluate the price elasticity of supply for ethanol. The average was 0.425 with all of the values less than 1.71, implying inelasticity. The price elasticity of demand for ethylene was estimated using limited data available for ethylene prices in the United States during 2008 to 2009. The value of -0.416 for the price elasticity of demand for ethylene was the result of a decrease in price that resulted in an increase in demand. All of these values imply inelasticity, which is probably typical for commodity chemicals.

Estimates of price elasticity of demand and supply can be incorporated as constraints in optimization of a chemical complex. Estimates of supply and demand changes will move upper and lower limits on availability of raw materials and demand for products. With price elasticity values, corresponding changes in prices are estimated for these changes in quantities of raw materials and products. Using price elasticity of demand and supply it was seen that for change in price of carbon dioxide, the demand for ammonia reduced. Cross price elasticity estimations were done to compute the PED of ammonia with respect to changes in carbon dioxide cost.

Data Used for Price Elasticity of Demand and Supply for Corn

Table B-9 Price Elasticity of Supply for Corn (USDA, 2010)

	Total Corn Supply				
Year	Production (million bushels per year)	DelQ/Q*100	Price ($/bushel)	DelP/P*100	PES
1981	8,675	9.65	3.11	-19.61	-0.49
1982	9,511	13.26	2.50	2.00	6.63
1983	10,772	-28.53	2.55	25.88	-1.10
1984	7,699	12.74	3.21	-18.07	-0.71
1985	8,680	21.35	2.63	-15.21	-1.40
1986	10,534	16.46	2.23	-32.74	-0.50
1987	12,267	-2.04	1.5	29.33	-0.07
1988	12,016	-23.52	1.94	30.93	-0.76
1989	9,191	2.98	2.54	-7.09	-0.42
1990	9,464	-1.93	2.36	-3.39	0.57
1991	9,282	-2.87	2.28	3.95	-0.73
1992	9,016	17.40	2.37	-12.66	-1.37
1993	10,584	-19.96	2.07	20.77	-0.96
1994	8,472	28.79	2.50	-9.60	-3.00
1995	10,910	-17.74	2.26	43.36	-0.41
1996	8,974	7.77	3.24	-16.36	-0.48
1997	9,672	4.42	2.71	-10.33	-0.43
1998	10,099	9.77	2.43	-20.16	-0.48
1999	11,085	1.33	1.94	-6.19	-0.21
2000	11,232	3.62	1.82	1.65	2.20
2001	11,639	-1.96	1.85	6.49	-0.30
2002	11,412	-7.31	1.97	17.77	-0.41
2003	10,578	5.77	2.32	4.31	1.34
2004	11,188	14.18	2.42	-14.88	-0.95
2005	12,775	3.60	2.06	-2.91	-1.24
2006	13,235	-5.48	2.00	52.00	-0.11
2007	12,510	14.80	3.04	38.16	0.39
2008	14,362	-4.40	4.20	-3.33	1.32
2009	13,729	7.75	4.06	-12.56	-0.62
2010	14,793	-0.41	3.55	5.63	-0.07
2011	14,733		3.75		
				Average PES	-0.16

Table B-10 Price Elasticity of Demand for Corn in Fuel Use (USDA (b), 2010)

Year	Demand(million bushels per year)	DelQ/Q*100	Price ($/bushel)	DelP/P*100	PED
	Corn used for fuel alcohol production				
1981	35.00	145.71	3.11	-19.61	-7.43
1982	86.00	62.79	2.50	2.00	31.40
1983	140.00	14.29	2.55	25.88	0.55
1984	160.00	45.00	3.21	-18.07	-2.49
1985	232.00	16.81	2.63	-15.21	-1.11
1986	271.00	7.01	2.23	-32.74	-0.21
1987	289.99	-3.74	1.5	29.33	-0.13
1988	279.15	2.97	1.94	30.93	0.10
1989	287.45	11.83	2.54	-7.09	-1.67
1990	321.45	8.59	2.36	-3.39	-2.53
1991	349.07	14.09	2.28	3.95	3.57
1992	398.26	6.84	2.37	-12.66	-0.54
1993	425.51	7.70	2.07	20.77	0.37
1994	458.26	16.26	2.50	-9.60	-1.69
1995	532.79	-25.73	2.26	43.36	-0.59
1996	395.68	8.35	3.24	-16.36	-0.51
1997	428.72	13.76	2.71	-10.33	-1.33
1998	487.73	6.17	2.43	-20.16	-0.31
1999	517.82	9.27	1.94	-6.19	-1.50
2000	565.85	11.31	1.82	1.65	6.86
2001	629.83	12.29	1.85	6.49	1.89
2002	707.24	40.76	1.97	17.77	2.29
2003	995.50	17.28	2.32	4.31	4.01
2004	1,167.55	13.33	2.42	-14.88	-0.90
2005	1,323.21	21.17	2.06	-2.91	-7.27
2006	1,603.32	32.19	2.00	52.00	0.62
2007	2,119.49	43.87	3.04	38.16	1.15
2008	3,049.21	20.58	4.20	-3.33	-6.18
2009	3,676.88	22.39	4.06	-12.56	-1.78
2010	4,500.00	4.44	3.55	5.63	0.79
2011	4,700.00		3.75		
				Average PED	0.51

Table B-11 Price Elasticity of Demand for Corn in Fuel Use (USDA, 2010)

Corn used for feed and residuals use					
Year	Demand(million bushels per year)	DelQ/Q*100	Price ($/bushel)	DelP/P*100	PED
1981	4,232	0.29	3.11	-19.61	-0.01
1982	4,245	7.74	2.50	2.00	3.87
1983	4,573	-15.24	2.55	25.88	-0.59
1984	3,876	6.15	3.21	-18.07	-0.34
1985	4,115	-0.01	2.63	-15.21	0.00
1986	4,114	13.25	2.23	-32.74	-0.40
1987	4,659	2.79	1.50	29.33	0.09
1988	4,789	-17.86	1.94	30.93	-0.58
1989	3,934	11.40	2.54	-7.09	-1.61
1990	4,382	5.17	2.36	-3.39	-1.52
1991	4,609	4.10	2.28	3.95	1.04
1992	4,798	9.47	2.37	-12.66	-0.75
1993	5,252	-10.90	2.07	20.77	-0.52
1994	4,680	16.66	2.50	-9.60	-1.74
1995	5,460	-14.05	2.26	43.36	-0.32
1996	4,692	12.46	3.24	-16.36	-0.76
1997	5,277	3.29	2.71	-10.33	-0.32
1998	5,450	0.04	2.43	-20.16	-0.00
1999	5,452	3.49	1.94	-6.19	-0.56
2000	5,643	3.18	1.82	1.65	1.93
2001	5,822	0.46	1.85	6.49	0.07
2002	5,849	-5.14	1.97	17.77	-0.29
2003	5,548	4.20	2.32	4.31	0.97
2004	5,781	6.12	2.42	-14.88	-0.41
2005	6,135	-0.33	2.06	-2.91	0.11
2006	6,115	-9.40	2.00	52.00	-0.18
2007	5,540	5.73	3.04	38.16	0.15
2008	5,858	-11.14	4.20	-3.33	3.34
2009	5,205	6.14	4.06	-12.56	-0.49
2010	5,525	-3.17	3.55	5.63	-0.56
2011	5,350		3.75		
				Average PED	-0.01

Appendix C Global Optimization Mathematics and Algorithms

In this appendix global optimization algorithms are describe for solving mixed-integer non-linear programming problems (MINLP) and multicriteria optimization problems such as the ones encountered in optimizing the chemical production complex. A description is given of global optimization methodology, solution techniques and GAMS solvers. A brief review is given of the various types of optimization problems and how they are solved. These include local optimization problems using linear programming, non-linear programming, mixed integer linear programming and mixed integer non-linear programming and global optimization problems using multicriteria or multi-objective optimization.

Global Optimization

Significant research has been spent developing algorithms that find the global optimum of a problem directly. This would eliminate using the procedure of finding all the local optima and then comparing these local optima to find the largest one the "global optimum".

Global optimization is the task of finding the absolutely best set of values of variables to optimize an objective function (Gray et al., 1997). Global optimization problems are typically difficult to solve. Global optimization problems are solved by extension of ideas from local optimization. These algorithms are integrated into computer programs for solving the problems. Both Pinter, 2014 and Trespalacios and Grossmann 2014 provide reviews of the more successful global algorithms and results of robustness *vs.* efficiency in practically motivated test problems,

The General Algebraic Modeling System (GAMS) is a high-level modeling language for mathematical programming and optimization. It consists of a language compiler and integrated high-performance solvers. GAMS is tailored for complex, large scale modeling applications, and allows building of large maintainable models that can be adapted quickly to new situations. The GAMS offers a wide range of solvers that allow the optimization based on type of problem. These include LP, NLP, MILP, MINLP and Global optimization solvers. The following section gives a summary of global optimization algorithms that have proved successful.

Global Optimization for Chemical Process Systems

Chemical process systems optimization problems frequently involve both continuous and binary variables and have the form of mixed integer nonlinear programming (MINLP) problems. The continuous variables represent the flow rates, temperature, pressures, etc., and binary variables represent the configuration of process units. These problems have been difficult to solve, and a significant amount of research has been spent developing algorithms that are effective in solving MINLP problems for the global optimum. The results have been improved algorithms (global optimizers) implemented in relatively reliable computer programs for both mixed integer linear and nonlinear programming problems. The mathematical form of a MINLP problem can be expressed as:

$$\text{Minimize:} \quad z = c^T y + f(x) \qquad \text{(C-1)}$$
$$\text{Subject to:} \quad Ay + h(x) = 0$$
$$By + g(x) \leq 0$$
$$x \in X = \{x | x \in R^n, x^L \leq x \leq x^U\}$$
$$y \in Y = \{y | y \in \{0,1\}^m, Ay \leq a\}$$

where **x** is a vector of continuous variables that represent the process variables such as flow rates, temperature, pressures, etc., and y is a set of binary variables that can be used to define the topology of the system representing the existence or non-existence of different processing units. The nonlinearities in the economic and process models appear in the terms $f(x)$, $g(x)$ and $h(x)$.

If any of the functions in Equations C-1 are non-linear, the problem corresponds to a mixed integer non-linear programming problem. If all functions are linear, it corresponds to a mixed-integer linear programming problem. If there are no binary variables (0-1) then the problem reduces to a non-linear programming problem or linear programming problem depending on whether the functions are nonlinear or linear. If there are only binary variables present, then it is an integer programing problem.

Global Optimization Algorithms

Global optimization is a branch of applied mathematics and numerical analysis that deals with the global optimization of a function or a set of functions according to some criteria. Typically, a set of bound and more general constraints is also present, and the decision variables are optimized considering also the constraints. Global optimization is distinguished from regular optimization by its focus on finding the maximum or minimum over all input values, as opposed to finding local minima or maxima. The *Journal of Global Optimization*, Springer, is one source of numerous publications on the multiplicity of methods tried to solve global optimization problems.

Global optimization algorithms are either deterministic or stochastic methods. The most successful deterministic strategies include inner and outer approximation methods, branch and bound methods, cutting plane methods and interval bounding methods. Successful stochastic strategies include random search, genetic algorithms and simulated annealing.

Inner and Outer Approximation Methods: Outer-approximation (OA) makes use of two main problems: (M-MIP) and (fx-MINLP). The main idea is to use the approximate linear problem (M-MIP) to find a lower bound of the objective function, (z^{lo}) (if maximizing) and obtain an integer solution to the approximate problem (y^P). This lower bounding problem is called master problem. For the subproblem, the binary variables (y^P) are fixed, and then (fx-MINLP) problem is solved. If the solution to (fx-MINLP) is feasible, then it provides an upper bound z^{up} (if maximizing). If it is not, (feas-MINLP) is solved to provide information about the subproblem, and an inequality that cuts off that integer solution is added. This method is performed iteratively until the gap of z^{lo} and z^{up} (the best upper bound) is less than the specified tolerance. At each iteration, the subproblem (either (fx-MINLP) or (feas-MINLP)) provides a solution (x^P, y^P), which is then included in the master problem (M-MIP) to improve the approximation. Since the function linearizations are accumulated, the lower bounding problem (or master problem) yields a nondecreasing lower bound ($z_1^{lo} \leq z_2^{lo} \leq \dots \leq z_p^{lo}$). A detailed description of this algorithm is given by Trespalacios and Grossmann, 2014 and Schaffer, 2012.

Branch and Bound Methods: These methods use a systematic enumeration of candidate solutions that are thought of as forming a tree with the full set of solutions at the top of the tree. The algorithm explores branches of this tree that represent subsets of the solution set. Each branch is checked against upper and lower estimated bounds on the optimal solution and ones that are discarded if they cannot produce a better solution than the best one found so far by the algorithm. The nonlinear branch and bound is an extension to this well known linear branch and bound. To find optimality, the method performs a tree search on the integer variables. It first solves the continuous relaxation (r-MINLP). If the solution yields integer values to all integer variables, then it is optimal and the algorithm stops. If it is not, a branching heuristic is to select an integer variable whose value at the current node is not integer ($y_i = y_i^0$). A branching is performed in this variable, giving rise to two new NLP problems. One NLP includes the bound $y_i \leq y_i^0$ while the other one $y_i \geq y_i^0$ or $y_i = 0$ or $y_i = 1$ if the integer variables are $0 - 1$ variables. This procedure is repeated until the tree search is exhausted. If an integer feasible solution is found, i.e., the solution provides integer values to all the integer variables, then it provides an upper bound. There are two cases in which some of the nodes are pruned, which makes the branch and bound method faster than enumerating every node. The first case in which a node is pruned occurs when the NLP corresponding to the node is infeasible. The second case occurs when the solution of the NLP of the node is larger than the current upper bound (Trespalacios and Grossmann 2014).

Generalized Benders Decomposition (GBD): This method is similar to the OA method, but they differ in the linear master problem. In particular, the master problem of the GBD only considers the discrete variables $y \in Y$, and the active inequalities $J^p = \{ j | g_j(x^p, y^p) = 0 \}$. Details for this algorithm are given by Trespalacios and Grossmann 2014.

Extended Cutting Plane (ECP): This method follows a similar concept to the OA method, but it avoids solving NLP sub-problems. In this method, at a given solution of the master MILP (M-MIP), all the constraints are linearized. A subset of the most violated linearized constraints is then added to the master problem. Convergence is achieved when the maximum violation lies within the specified tolerance. The algorithm provides a nondecreasing a lower bound after each iteration. The main strength of the method is that it relies solely in the solution of MILPs, for which powerful algorithms are readily available. Similarly to the OA method, it solves the problem in one iteration if $f(x, y)$, and $g(x, y)$ are linear. There are two main downsides in the algorithm. The first one is that convergence can be slow, since the convergence in the continuous space follows from the convergence of Kelley's cutting-plane method, which is in general linear. The second one is that the algorithm does not provide an upper bound (or feasible solution) until it converges (Trespalacios and Grossmann 2014).

Interval Methods: These methods start by bounding the intervals on the independent variables that contain the global optimum. Then they proceed to reduce the bounds on these variables by various means to have final intervals of the desired precision containing the global optimum. These types of methods evaluate each constraint with the current variable bounds, and try to improve bounds by maintaining feasibility in the constraint. A recent method uses pairs of constraints instead of individual constraints to infer bounds. Different techniques have been developed to infer bounds on MILP problems and on MINLP problems. Details are provided by Trespalacios and Grossmann 2014.

Stochastic Methods: The more successful stochastic strategies include random search, genetic algorithms and simulated annealing. Random search is a stochastic method that places measurements (evaluation of the objective function) randomly in the initial intervals of the independent variables. Depending on the number of experiments used, the values of the objective function are ranked, and it can be said statistically that the maximum (or minimum) is in the top x percent with a y probability. The values of the initial intervals can be adjusted based on these results to have smaller region to search, and random measurements are placed in the new region (creeping random search). See Pike, 2013.

Genetic algorithms, annealing algorithms, tabu search, artificial neural networks, among others, use randomized search techniques for finding near optimal solutions of combinatorial optimization problems (Pardalos and Resende, 2002 and Schaffer, 2012). The idea behind using artificial neural networks is to map the optimization problem into a highly interconnected network of neurons, and a particular configuration of neurons being on or off determines the value of the objective function. The procedure uses an activation function to transform the neurons to locate the configuration that approaches the global solution of the objective function. A sigmoid function is said to be the most used activation function in the artificial neural network literature (Trafalis and Kaspa 2002.)

Simulated annealing is a family of randomized algorithms for locating near optimal solutions of combinatorial optimization problems using the idea of annealing in metallurgy, a technique involving heating and controlled cooling of a material to increase the size of its crystals and reduce their defects. Slow cooling is used as an analogy to decrease in the probability of accepting worse solutions as it explores the solution space because it allows for a more extensive search for the optimal solution. Steps with improvements are accepted and ones that do not improve the value of the objective function are accepted within a certain probability. The goal is to bring the system, from an arbitrary initial state to a state with the minimum possible thermodynamic free energy. Threshold algorithms are used to move to improved values of the objective function and are described by Aarts and Ten Eikelder 2002.

Genetic algorithms (Goldberg, 1989) use search heuristic that mimics the process of natural selection to generate useful solutions to optimization problems. The initial solution starts from a population of randomly generated individuals and moves based on heuristics. New solutions are combined with old solutions to generate improved solutions, ones that move to the optimum of the objective function. The algorithm terminates when a maximum number of iterations is reached or a satisfactory value of the objective function has been obtained.

Summary: An overview of the start-of-the-art in software for the solution of mixed integer nonlinear programs (MINLP) is given by Bussieck and Vigerske, 2014, of GAMS. They establish several groupings with respect to various features and give concise individual descriptions for each solver. The objective is to provided information to guide the selection of a best solver for a particular MINLP problem. Global optimization of MINLP requires an effective algorithm or combination of algorithms, usually LP, MIP and NLP, implemented in programming languages, and run on a computer with an operating system for linear or parallel operations. Over time there have been research results reported on efficient algorithms for sets of problems, and the sets of problems have become comparable in size to industrial plants and algorithms (solvers) have improved correspondingly. Comparisons of global optimization programs (solvers) are given subsequently for a chemical production complex optimization with new processes for chemicals from biomass (Sengupta and Pike, 2012.)

GAMS (General Algebraic Modeling System) Programming Language

The GAMS (General Algebraic Modeling System) programming language was developed by the GAMS Development Corporation 1217 Potomac Street, NW, Washington, D.C. 20007 (http://www.gams.com). GAMS is a high-level modeling language for mathematical programming and optimization. It consists of a language compiler and a stable of integrated high-performance solvers. GAMS is tailored for complex, large scale modeling applications, and large maintainable models can be adapted quickly to new situations.

GAMS is specifically designed for solving linear, nonlinear and mixed integer optimization problems. The system is especially useful with large, complex problems. GAMS is available for use on personal computers, workstations, mainframes and supercomputers. GAMS is able to formulate models in many different types of problem classes, and switching from one model type to another can be done with a minimum of effort. The same data, variables, and equations can be used in different types of models at the same time.

GAMS model types include Linear Programming (LP), Mixed-Integer Programming (MIP), Mixed-Integer Non-Linear Programming (MINLP), and different forms of Non-Linear Programming (NLP). There are over 30 solvers (optimization codes) that can be selected to solve these programming problems. Note, "programming," means "scheduling" and not "computer programming."

GAMS Distribution 24.3.3 is currently available (11-10-2014) for download from the GAMS web site www.GAMS.com without charge. GAMS will operate as a free demo system without a valid GAMS license. The model limits in demo mode are 300 constraints and variables, 2000 nonzero elements, (of which 1000 can be nonlinear), 50 discrete variables (including semi continuous, semi integer and member of SOS-Sets) with additional global solver limits of 10 constraints and variables. There are the installation notes for Windows, Mac, and UNIX. The GAMS distribution includes the GAMS Manuals in electronic form, and hard copies can be ordered through Amazon.

GAMS uses solvers developed for the above optimization algorithms for various types of problems. An extensive list of solvers can be found at GAMS website (www.GAMS.com) for solving LP, NLP, MIP, MILP and MINLP problems. The solvers used to solve the global optimization problem in the Chemical Complex Analysis System were BARON and LINDOGLOBA.

178

The NEOS Server for Optimization hosted by the Argonne National Laboratory is an open and free to use server for solving optimization problems (NEOS, 2010). The optimization solvers at NEOS represent the state-of-the-art in optimization software. Optimization problems are solved automatically with minimal input from the user. The users only need a definition of the optimization problem, and all additional information required by the optimization solver is determined automatically by the server. For example, the solver choice for MINLP is required, but the sub-choices for LP and NLP need not be specified in the server.

The optimal structure was determined from the superstructure of global optimization problem in the Chemical Complex Analysis System using five different solvers from the NEOS server. These were DICOPT, SBB, BARON, ALPHAECP and LINDOGLOBAL. Two of these solvers were listed exclusively under global solvers that accepted GAMS input (BARON and LINDOGLOBAL), and the other three were listed under MINLP solvers (DICOPT, SBB, ALPHAECP). The results for computation time and solver status from the NEOS server solution are given in Table C-1 from Sengupta and Pike, 2012. The SBB, DICOPT and BARON gave a normal completion with identical solutions for the objective value. Computational, generation and execution times were comparable. The LINDOGLOBAL was unable to solve because of an iteration interrupt. The ALPHAECP gave a normal completion with infeasible solution. Table C-2 gives the comparison of the solution using SBB in the NEOS server and the local machine, an Intel PC; and the results were the same.

Multiobjective Optimization

Multiobjective optimization, also called multicriteria optimization, is the simultaneous optimization of more than one objective function. The general Multiobjective Problem (MOP) is defined as in Equation C-2:

Optimize: $\quad F(\mathbf{x}) = [f_1(\mathbf{x}), f_2(\mathbf{x}), \ldots, f_k(\mathbf{x})]^T$

$$\text{(C-2)}$$

Subject to: $\quad g_i(\mathbf{x}) \geq 0 \qquad i = 1, 2, \ldots, m$
$\qquad\qquad\quad h_j(\mathbf{x}) = 0 \qquad j = 1, 2, \ldots, p$
$\qquad\qquad\qquad\qquad\quad x^L \leq x \leq x^U$

There are various methods to solve multicriteria optimization problems like utility function, hierarchical methods and goal programming (Rangaiah and Bonilla-Petriciolet 2013 and Rao, 2009). Of these, using the utility function or weighted objective method is the most commonly used. In this method, weights are assigned to the different objective functions and the sum of the weights times the objective functions is formed for a single objective function as shown in Equation C-2. The sum of the weights can equal 1.0. This is represented in Equation C-2.

Optimize: $\quad F'(\mathbf{x}) = \sum_{i=1}^{n} w_i f_i(\mathbf{x})$

$$\text{(C-1)}$$

$$\sum_{i=1}^{n} w_i = 1$$

Table C-1 Comparison of Solvers in NEOS Server for Optimal Solution

Solver	SBB (MINLP)	DICOPT (MINLP)	ALPHAECP (MINLP)	BARON (Global)	LINDOGLOBAL (Global)
OBJECTIVE VALUE	16.500316	16.500313	NA	16.49418566	NA
SOLVER STATUS	NORMAL COMPLETION	NORMAL COMPLETION	NORMAL COMPLETION	NORMAL COMPLETION	ITERATION INTERRUPT
MODEL STATUS	INTEGER SOLUTION	INTEGER SOLUTION	INFEASIBLE - NO SOLUTION	INTEGER SOLUTION	NO SOLUTION RETURNED
Additional Solvers chosen by NEOS	CONOPT 3 (NLP)	XPRESS (MIP) CONOPT 3 (NLP)	-	ILOG CPLEX (LP) MINOS (NLP)	-
Iteration Count	246/10000	318/10000	47/10000	0/10000	0/10000
Resource Usage	0.340/1000.000	0.370/1000.000	62.110/1000.000	40.000/1000.000	10.336/1000.000
Compilation Time	0.037 SECONDS	0.034 SECONDS	0.036 SECONDS	0.034 SECONDS	0.037 SECONDS
Generation Time	0.024 SECONDS	0.025 SECONDS	0.014 SECONDS	0.025 SECONDS	0.014 SECONDS
Execution Time	0.026 SECONDS	0.027 SECONDS	0.016 SECONDS	0.027 SECONDS	0.016 SECONDS

Table C-2 Comparison of Solvers in NEOS Server and Local Machine

Solver	SBB (MINLP) (NEOS Server)	SBB (MINLP) (Local Machine)
OBJECTIVE VALUE	16.500316	16.500316
SOLVER STATUS	NORMAL COMPLETION	NORMAL COMPLETION
MODEL STATUS	INTEGER SOLUTION	INTEGER SOLUTION
GAMS version	GAMS Rev 228 x86/Linux	GAMS Rev 232 WIN-VIS 23.2.1 x86/MS Windows
Additional Solvers chosen by NEOS	CONOPT 3 (NLP)	CONOPT
Iteration Count	246/10000	214/ 2000000000
Resource Usage	0.340/1000.000	0.359/1000.000
Compilation Time	0.037 SECONDS	0.015 SECONDS
Generation Time	0.024 SECONDS	0.063 SECONDS
Execution Time	0.026 SECONDS	0.063 SECONDS

The multicriteria problem can be a mixed integer nonlinear programming problem where the multiple objective functions and the constraints are non-linear and the variables are continuous or integer. The MINLP problem in this research was formulated into a multi-criteria problem by maximizing the profit and the sustainability credits simultaneously.

A detailed review of multicriteria optimization in sustainable energy decision-making was given by Wang et al., 2009. Technical criteria, economic criteria, environmental criteria and social criteria were discussed in the paper along with weighted objective methods.

Multiobjective Optimization Problem Statement for a Chemical Production Complex

The statement for the optimization problem for a chemical production complex is:

Optimize: Objective Function
Subject to: Constraints from plant models

The objective function is a profit function for the triple bottom line, Equation 50.

$$\text{Triple Bottom Line} = \text{Profit} - \Sigma \text{ Environmental Costs} + \Sigma \text{ Sustainable (Credits} - \text{Costs)} \quad (50)$$

The profit is described using an extended value added economic model, Equation 49.

$$\text{Profit} = \Sigma \text{ Product Sales} - \Sigma \text{ Raw Material Costs} - \Sigma \text{ Energy Costs} \quad (49)$$

$$\text{Triple Bottom Line} = \Sigma \text{ Product Sales} - \Sigma \text{ Raw Material Costs} - \Sigma \text{ Energy Costs} -$$
$$\Sigma \text{ Environmental Costs} + \Sigma \text{ Sustainable (Credits} - \text{Costs)}$$

The constraint equations describe relationship among variables and parameters in the processes and plants. Equality constraints are material and energy balances, chemical reaction rates, thermodynamic equilibrium relations and others. Inequality constraints are availability of raw materials, demand for products, capacities of process units and others.

The objective of multicriteria optimization is to find optimal solutions that maximize industry' profits and minimize costs to society. This multicriteria optimization problem can be stated as in terms of industry's profit, P, and society's sustainable credits/costs, S; and these two objectives are given by Equation 51.

Max: $P = \Sigma$ Product Sales - Σ Economic Costs - Σ Environmental Costs   (51)
          $S = \Sigma$ Sustainable (Credits $-$ Costs)

Subject to: Multi-plant material and energy balances,
                   product demand, raw material availability, plant capacities

To locate Pareto optimal solutions, multi-criteria optimization problems are converted a single criterion by applying weights to each objective and optimizing the sum of the weighted objectives as shown in Equation 52 where $w_1 + w_2 = 1$.

Max: $w_1 P + w_2 S = w_1 P + (1 - w_1) S$ (52)

Subject to: Multi-plant material and energy balances,
             product demand, raw material availability, plant capacities

If w_1 is 0, then only industry profits are considered and no sustainable costs/credits are included. If $w_1 = 1$ the only sustainable costs/credits are evaluated at the Pareto optimum. With $w_1 = 0.5$ industry profits and sustainable cost/credits are weighted equally. Results are summarized in Figure 35 for the chemical production complex. It is another decision to determine the specific value of the weight that is acceptable to all concerned.

GAMS (General Algebraic Modeling System) was used in the Chemical Complex Analysis System to determine the optimum configuration of plants based on economic, environmental and sustainable costs reported in Figure 34. Optimal solutions generated by multicriteria optimization are shown in Figure 35 for a range of values for w_1 from 0 to 1.0.

Notes on Algorithms for Local Optimization

Linear Programming (LP) is the simplest type of optimization problem where the objective function and constraint equations are all linear, and it is the most widely applied of all of the optimization methods. The technique has been used for optimizing many diverse applications. The general equation for LP can be written as given in Equation B-7. Linear programming requires all constraint equations be written as equalities. Inequalities need to be converted to equality constraints using slack and surplus variables.

Optimize: $\displaystyle\sum_{j=1}^{n} c_j x_j$

                                                                                        (C-2)

Subject to: $\displaystyle\sum_{j=1}^{n} a_{ij} x_j = b_i$

             $x_j \geq 0$ $i = 1, 2, \ldots m \quad j = 1, 2, \ldots n,$

Linear programming problems are usually solved with computer programs that use the revised simplex method. See Arora, 2012 and Rao, 2009 for details.

Non Linear Programming (NLP) refers to multivariable optimization procedures where the equation for objective function and constraint equations are non-linear functions of variables. The general representation of the NLP problem is given as in Equation C-4. There are n independent variables, $x = (x_1, x_2, \ldots, x_n)$, m constraint equations, h of them being equality constraints. The values of x_j's can have lower and upper bounds specified.

Optimize: y($\mathbf{x}$)

(C-3)

Subject to: $f_i(\mathbf{x}) = 0$ for i = 1,2,....,h
               $f_i(\mathbf{x}) \geq 0$ for i = h+1,....,m

Non-linear programming problems are usually solved with computer programs that use the generalized reduced gradient algorithm or the sequential quadratic programming algorithm. See Pike, 2013 for details.

Mixed Integer Linear Programming (MILP) refers to an extension of linear programming problem where some variables are required to be integers. The use of integer variables makes possible the formulation of models with a discrete selection of variables or constraints. The problem statement of MILP is given in Equation C-5. When all the variables are integers, the problem is referred to as Integer Programming (IP). A special case of Integer Programming is Binary Integer Programming (BIP), where variables take only 1 or 0 as values. BIP is applied to problems where "yes-or-no decisions" are important.

Optimize: $\mathbf{c}^T\mathbf{x} + \mathbf{h}^T\mathbf{y}$

(C-4)

Subject to: $\mathbf{Ax} + \mathbf{Gy} \leq \mathbf{b}$

$\mathbf{A}$ and $\mathbf{G}$ are m × n and m × p matrices;
$\mathbf{b}$ is m-dimensional vector
$\mathbf{x}^T = (x_1,\ldots x_n)$ $\mathbf{x}$ is n dimensional vector of positivecontinuous variables
$\mathbf{y}^T = (y_1,\ldots y_p)$ $\mathbf{y}$ is p-dimensional vector of positive integer variables

Mixed Integer programming problems are usually solved with computer programs that use the branch and bound algorithm. See Ecker, J. G. and M. Kupferschmid, 1988 and Murty, K. G., 1976 for details.

Summary

Global optimization algorithms have been describe for solving mixed-integer non-linear programming problems (MINLP) and multicriteria optimization problems such as the ones encountered in optimizing the chemical production complex. The most successful deterministic strategies for global optimization algorithms included inner and outer approximation methods and branch and bound methods.

The GAMS (General Algebraic Modeling System) programming language was specifically designed for solving large, complex linear, nonlinear and mixed integer optimization problems. GAMS was used to determine the optimal structure of global optimization problem in the Chemical Complex Analysis System with five different solvers. Two of these solvers were listed exclusively under global solvers that accepted GAMS input (BARON and

LINDOGLOBAL), and the other three were listed under MINLP solvers (DICOPT, SBB, ALPHAECP).

GAMS was used in the Chemical Complex Analysis System to determine the optimal solutions generated by multicriteria optimization. Results are shown in Figure 35 for a range of values for w_1 from 0 to 1.0.

Monte Carlo simulation was used to determine the sensitivity of the optimal solution to the costs and prices used in the triple bottom line. One of the results was the cumulative probability distribution, a curve of the probability as a function of the triple bottom line, that was used to determine upside and downside risks, Figure 36.

References

Aarts, E. H. L. and H. M. M. Ten Eikelder, 2002, " Simulated Annealing," *Handbook of Applied Optimization*, Pardalos, P. M. and M. G. C. Resende, Editors, Oxford University Press, New York, NY
Arora, J. S. 2012, *Introduction to Optimum Design*, 3rd Ed., Academic Press, Waltham, MA

Ecker, J. G. and M. Kupferschmid, 1988 *Introduction of Operations Research*, Wiley, New York.

Goldberg, D. E., 1989, *Genetic Algorithms in Search, Optimization and Machine Learning*, Addison –Wesley, New York

Gray, P. W. Hart, L. Painton, C. Phillips, M. Trahan, J. Wagner, *A Survey of Global Optimization Methods, Sandia* National Laboratories, Albuquerque, NM 87185
www.cs.sandia.gov/opt/**survey**/ accessed 11/11/2014

Grossmann I, Caballero JA, Yeomans H (1999); Mathematical Programming Approaches to the Synthesis of Chemical Process Systems, Korean J. Chem. Eng., 16(4), 407-426.

Murty, K. G., 1976 *Linear and Combinatorial Programming*, Wiley, New York.
NEOS (2010); NEOS Server for Optimization, http://www-neos.mcs.anl.gov/

Pardalos, P. M. and M. G. C. Resende, Editors, 2002, *Handbook of Applied Optimization*, Oxford University Press, New York, NY

Pike, Ralph W. 2013, *Optimization for Engineering Systems Revised*, (Kindle Edition) ASIN: B00BF2TLXO Amazon.com (2013)

Pinter, J. D., 2014, "How Difficult is Nonlinear Optimization? A Practical Solver Tuning Approach, with Illustrative Results " *Optimization Online*, Mathematical Optimization Society, June, 2014

Rangaiah and Bonilla-Petriciolet 2013, *Multiobjective Optimization in Chemical Engineering: Developments and Applications*, Wiley, Hoboken, NJ

Rao, 2009, *Engineering Optimization Theory and Practice*, Fourth Ed., Wiley, Hoboken, NJ

Schaffer, Stefan, 2012, *Global Optimization: A Stochastic Approach*, Springer, New York, NY

Sengupta, Debalina and Ralph W. Pike,2012, *Chemicals from Biomass: Integrating Bioprocesses into Chemical Production Complexes for Sustainable Development*, CRC Press, Boca Raton, FL

Trafalis, T. B. and S. Kaspa 2002, "Artificial Neural Networks in Optimization and Applications," *Handbook of Applied Optimization*, Pardalos, P. M. and M. G. C. Resende, Editors, Oxford University Press, New York, NY

Trespalacios, F. and I. E. Grossmann 2014, "Review of Mixed-Integer Nonlinear and Generalized Disjunctive Programming Methods," *Chem. Ing. Tech.*, Vol. 86, No. 7, p. 991–1012

Wang JJ, Jing YY, Zhang CF, Zhao JH (2009); Review on multi-criteria decision analysis aid in sustainable energy decision-making, *Renewable and Sustainable Energy Reviews*, 13(9): 2263-2278.

About the Author

Ralph W. Pike is the Director of the Minerals Processing Research Division and is the Paul M. Horton Professor Emeritus of Chemical Engineering at Louisiana State University. His degrees are in chemical engineering, doctorate and bachelors from the Georgia Institute of Technology. He is the author of a textbook on optimization for engineering systems and coauthor of four other books on design and modeling of chemical processes. He has directed fifteen doctoral dissertations and sixteen masters' theses in chemical engineering. He is a registered professional engineer in Louisiana and Texas. His research has been sponsored by Federal and State agencies and private organizations with 107 awards totaling $5.6 million and has resulted in over 200 publications and presentations. His research specialties are optimization theory and applications for the optimal design of engineering systems, on-line optimization of continuous processes, optimization of chemical production complexes, and related areas of resources management, sustainable development, continuous processes for carbon nanotubes and chemicals from biomass.

He is a Fellow of the American Institute of Chemical Engineers (AIChE) and is the past-chair of the Environmental Division and of the Fuels and Petrochemicals Division. He is an active member of the Institute for Sustainability and the Safety and Chemical Engineering Education (SACHE) committee of the Center for Chemical Process Safety. He was the Meeting Program Chairman for 74th Annual Meeting and has co-chaired 66 sessions on optimization, sustainability, transport phenomena, reaction engineering. He has held all of the positions in the Baton Rouge Section of the AIChE. He is on the Editorial Boards of the journal *Environmental Progress and Renewable Energy* and the journal *Clean Technology and Environmental Policy*. He has served as Co-Editor-in-Chief of *Waste Management*, an international journal devoted to information on prevention, control, detoxification and disposition of hazardous, radioactive and industrial wastes. He is a member of the American Chemical Society and Sigma Xi, the scientific society.

Other books that he has authored and co-authored:

Ralph W. Pike, *Optimization for Engineering Systems Revised,* (Kindle Edition) ASIN: B00BF2TLXO Amazon.com (2013).

Debalina Sengupta and Ralph W. Pike, *Chemicals from Biomass: Integrating Bioprocesses into Chemical Production Complexes for Sustainable Development*, CRC Press, Boca Raton, FL, 2012.

Richard C. Farmer, Ralph W. Pike, Gary C. Cheng, and Yen-Sen Chen, *Computational Transport Phenomena for Engineering Analysis*, Taylor & Francis Publishing Company, 2009.